university college
for the creative arts

at canterbury, epsom, farnham
maidstone and rochester

Stereoscopic Cinema and the Origins of 3-D Film

Stereoscopic Cinema & the Origins of 3-D Film, 1838–1952

Ray Zone

THE UNIVERSITY PRESS OF KENTUCKY

Ray Zone is an award-winning 3-D artist, author, and speaker whose articles about 3-D cinema have appeared in the *Los Angeles Times, American Cinematographer,* and *Hollywood Reporter.* A 3-D film producer, Zone has produced or published over a hundred 3-D comic books and is the author of *3-D Filmmakers: Conversations with Creators of Stereoscopic Motion Pictures* (2005). Zone's website is viewable in anaglyphic 3-D and is at www.ray3dzone.com.

Publication of this volume was made possible in part by a grant from the National Endowment for the Humanities.

© 2007 by The University Press of Kentucky
Scholarly publisher for the Commonwealth, serving Bellarmine University,
Berea College, Centre College of Kentucky, Eastern Kentucky University,
The Filson Historical Society, Georgetown College, Kentucky Historical Society,
Kentucky State University, Morehead State University, Murray State University,
Northern Kentucky University, Transylvania University, University of Kentucky,
University of Louisville, and Western Kentucky University.

Editorial and Sales Offices: The University Press of Kentucky
663 South Limestone Street, Lexington, Kentucky 40508-4008
www.kentuckypress.com

Frontispiece: March 1925 ad from *Motion Picture News.*

11 10 09 08 07   5 4 3 2 1

Library of Congress Cataloging-in-Publication Data

Zone, Ray.
  Stereoscopic cinema and the origins of 3-D film, 1838–1952 / Ray Zone.
     p. cm.
  Includes bibliographical references and index.
  ISBN 978-0-8131-2461-2 (hardcover : alk. paper)
  1. 3-D films—History. I. Title.
  TR854.Z68 2007
  778.5'34109—dc22   2007012691

This book is printed on acid-free recycled paper meeting the requirements of the American National Standard for Permanence in Paper for Printed Library Materials.

Manufactured in the United States of America.

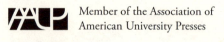

Member of the Association of
American University Presses

Contents

Foreword

IT WAS CHRISTMAS VACATION IN 1952. The snow was falling in Rockefeller Plaza in Manhattan, but I was warm and dry as I stood in the queue with Harvey, Morty, Jeffrey, and Robby, waiting to see ourselves on closed-circuit color TV in the storefront RCA Exhibition Hall. We inched along, mingling with the holiday crowd, until we came into the field of view of the lens of the refrigerator-sized color camera and at last saw ourselves on a nearly circular color tube, in bright "living colors." Only there were too many living colors. The picture was out of alignment or, more properly, convergence. No such shortcoming could dim my enthusiasm, but when I got home to tell my mom how much I wanted a color TV, she wisely said: "We'll buy one once it's perfected." It turns out we never had a color TV in our living room (I bought her one years later), and neither did our Brooklyn neighbors, for it took more than twenty years and a fortune in research and development and marketing before the Radio Corporation of America (RCA) and its licensees achieved a 50 percent penetration for color TV sets. By that time, the image quality had improved, but the system, named after a standards body empowered by RCA, was mocked in engineering circles, and NTSC (National Television Standards Committee) came to be known as Never Twice the Same Color.

As I read the manuscript of Ray Zone's technology history, I thought of that day seemingly a lifetime ago, when my friends and I waited to catch a glimpse of ourselves, and I think about the adoption of technology in my lifetime. Today, practically all of us are using computers—which are still an unfinished piece of business. An automobile with the characteristics of a personal computer would be classified as an unreliable and dangerous instrument. Fortunately, computer crashes are more frequent than car crashes.

So what does it take for us to accept a new technology? Ray Zone's history digs at that question and intertwines the efforts of inventors who sought to create the motion picture with those who sought to create the stereoscopic motion picture (often the same inventor), and he makes the case that the development of stereoscopic technologies strongly influenced the movies. He makes the case with scores of examples, and what is so interesting about this is that other writers have not noticed this connection. The stereoscopic cinema, like early color television, is not held in high esteem. It is more frequently derided as an eye-straining fad than praised for its potential contribution to the art of the cinema. This may explain why no one else has made Ray's connection. That's because writers like Ray, who have a measure of respect for the three-dimensional (3-D) cinema, are few and far between. Most lack the vision.

I have spent most of my life as an inventor of stereoscopic moving image systems. Lately my work has taken me to the theatrical cinema, although it was more than twenty years ago that I made the basic invention that led to the latest renaissance in 3-D digital movies. I was aware of some of the history of the medium, and I knew something about the men who put their time and energy—and to tell the truth in many cases, their life's blood—into the invention of the cinema and stereoscopic displays. But Ray knows a lot more of the story.

It is a history replete with many false starts and often of people working ceaselessly only to have their inventions die a stillbirth. It is a history that could not have made any sense in the moment and can be seen to have a form only with the passage of time and the boon of hindsight. Ray has presented the facts of the early days of the parallel development of cinema and stereoscopic display technology and to some extent left it to us to connect the dots. It is often difficult to say how one inventor influenced another. Were they aware of the work of their contemporaries? How many of the many inventions described in this book got built? And how many that got built actually worked? Ray has comments about this and relies wherever possible on eyewitness reports, but these are often inexpert and contradictory, which is what one would expect in historical research.

In my view, the way to best trace the milestones is to connect the dots between actual products. Such an approach involves emphasizing devices that were offered in the marketplace with at least a modicum of commercial success—not because the ability to monetize an invention is necessarily a measure of its worth but, rather, because it's a measure of how many people saw it in use. That's because the inventors are in the theaters along with the rest of the public, and a visual experience—like mine at the RCA Exhibition Hall—is more profound and influential than reading a paper or a patent.

This introducer is not out of things to say about Ray Zone's book, but I am just about out of my allotment of words. As a professional inventor in this field, it is only natural that I would feel that this book was meant for me, but I think it is a book that will be important for years to come to both enthusiasts and students of the technology of the cinema and stereoscopy.

—Lenny Lipton
Laurel Canyon, California

Lenny Lipton is the inventor of the field sequential technology at the heart of digital 3-D cinema, CTO of RealD, and the author of numerous books on stereoscopic television and cinema.

Acknowledgments

I AM GREATLY INDEBTED TO Marvin Jones, John Dennis, and Jan Burandt, editors of the Stereo Club of Southern California's *3D News, Stereo World,* and *Stereoscopy* magazines, respectively, in which portions of this book first appeared.

The research assistance of David Burder, Arthur Girling, Paul Wing, Gordon Trewinnard, Dieter Lorenz, Stephen Herbert of the Projection Box, Alexander Klein of Stereoscopy.com, and collector/scholar Erkki Huhtamo has been invaluable.

Folks at the Margaret Herrick Library of the Academy of Motion Picture Arts and Sciences were very helpful. Special thanks go to Patricia Williams of the Los Angeles Public Library.

Many thanks to Monica Dunlop, Michael Georgoff, Jack Rutberg, Christian Bible, and Larry Cuba, who assisted with materials on the stereo avant-garde.

Ed Shaw, raconteur and great friend of Paul Wing, laid a packet of historical materials on the author in 2005 that factor large in the epilogue to this book. Thank you, Ed.

Dan Symmes, the 3D man, has proved tremendously helpful, and his hard work, along with that of Jeff Joseph and Bob Furmanek, at the World 3D Film Expos (I and II), has been a genuine inspiration.

My editor, Janet Foster, has done a superlatively Teutonic job of keeping me verbally trim and thematically focused. The passion of Lawrence Kaufman for 3-D movies has also been a guiding light for this often-beleaguered historian.

Thank you, Lenny Lipton, for your amazing stereographic career, great discussions about 3-D, and the beautiful foreword.

I could not have written this book without the love and assistance of Susan Pinsky and David Starkman, eminent stereographers and proprietors of Reel 3-D Enterprises, to whom this work is dedicated.

Prologue

It is Mr. Edison's intention to give a stereoscopic effect to the pictures taken in connection with the Kinetograph, and a long extensive series of experiments have been conducted at the Laboratory, very good results being obtained. This has all been incorporated in the patent.
—William Kennedy-Laurie Dickson, in a letter of June 16, 1891

HISTORICALLY, stereoscopic cinema can be divided into four general periods through which the "grammar" of stereographic narrative has evolved within the overall arena of cinema itself.

The Novelty Period—1838 to 1952

Whereas the "novelty period" for conventional cinema lasted about ten years (1895–1905), with stereoscopic cinema it continued for more than a century after the discovery of 3-D by Charles Wheatstone in 1838. This was due, in part, to a bewildering variety of technological strategies for creating a stereographic motion picture, as well as the continuous technical evolution of the conventional motion picture medium itself.

Though there were very few 3-D feature films produced during this period—among them, *The Power of Love* in anaglyph (1922), *M.A.R.S. or Radio-Mania* in the Teleview alternating frame system (1922), and *Robinson Crusoe*, produced in Russia by Semyon Ivanov (1941)—the short stereoscopic films of the novelty period are characterized by an emphasis on the technology of 3-D or the "gimmick" of the off-the-screen imagery that is evident in Ives-Leventhal's *Plastigrams* of the 1920s and *Audioscopiks*, Pete Smith's MGM short stereo films of the mid 1930s and early 1940s.

It was the novelty of the stereoscopic medium that was foregrounded in 3-D films of this period. During this period of technological progress for the stereoscopic motion picture, there was an aesthetic tug-of-war between the technical and the narrative demands of the medium.

By the end of stereoscopic cinema's novelty period, technical fundamentals had become widely understood, polarizing image selection was in use, the twin-strip 3-D films of John Norling had been seen by several million people at the 1939–1940 New York World's Fair, and the 1951 Festival of Britain had demonstrated artistic possibilities for the stereographic film.

An Era of Convergence—1952 to 1985

The stage had been set for the brief but protean 3-D movie boom of 1952–1955 in Hollywood, which began with Arch Oboler's *Bwana Devil*. The 1950s initiated the second era for stereoscopic motion pictures, which I have identified according to one of its optical and production characteristics, "convergence."

Imitating the movement of the human eyes, the 3-D motion picture cameras of the 1950s converged on subject matter, whether near or distant, and set the action within the confines of the conventional motion picture frame of the time (1:33 to 1, or Academy aperture). It was on this canvas that classical Hollywood had painted the movie epics of its golden era from 1939 to 1951. This same canvas was used by the stereographic storytellers during the 3-D boom of the 1950s.

With more than fifty stereoscopic films released between 1952 and 1955, William K. L. Dickson's vision of a crowning realism became a motion picture reality. Cinematic storytellers employed with varying results the expanded narrative palette that the stereographic motion picture presented. The stereoscopic films of the 1950s, using the Academy aperture and converging binocular optical axes, created awareness of the stereo window bounded by edges, a floating window on another apparent reality. When the exhibitors and studios elected to pursue Cinemascope with its (2:55 to 1) wider screen, the classical Hollywood frame was broken, along with the 3-D film.

The brief flurry of stereoscopic films in the early 1980s, beginning with

Comin' at Ya and continuing up to the summer of 1983 with *Jaws 3-D,* saw no great enlargement of the grammar of stereoscopic narrative. Exhibited in alternating-frame single-strip systems, or in anamorphically squeezed, side-by-side stereo pairs, the 1980s 3-D films often used a 1:85 to 1 aspect ratio, but the reliance on optical convergence and an unrelenting exploitation of off-the-screen effects disconnected from the narrative demonstrated only slight artistic advance for stereographic film.

The first real day-and-date wide release of a stereoscopic motion picture took place on August 12, 1982, with *Friday the 13th Part III* from Paramount Pictures opening on over 1,000 silver screens in North America.

The Immersive Era—1986 to Present

Technological innovation with large-format (LF) 15/70-mm IMAX film in the 1970s led to the next era of stereoscopic cinema history. This large-format 70-mm film travels horizontally through the camera and projector so that each frame is fifteen perforations wide, almost ten times the size of a conventional 35-mm film frame. Inauguration of the 15/70-mm IMAX 3-D format led to the third epoch for stereoscopic cinema.

With its seven-story-high image projected on the giant screen accompanied by six-channel sound, the large-format stereoscopic film eliminates the awareness of the edge of the frame as with the stereo window of the 1950s. The viewer's peripheral vision is completely immersed in the stereoscopic image, and there is no awareness of the edge of the motion picture frame. With LF 3-D, the stereoscopic imagery floats off the giant screen and into the audience space. Convergence of camera axes is rarely if ever used in 15/70 3-D, so that the stereo window is placed at infinity to bring the screen imagery entirely out into the theater.

The film that announced the "immersive" era of stereoscopic cinema and the giant screen was *Transitions,* a 1986 IMAX 3-D film produced by Colin Low for Vancouver Expo. More than thirty large-format 3-D films have been produced since 1986, and 115 permanent IMAX 3-D theaters have been constructed worldwide with seven-story-high silver screens and twin-strip "single-body" projectors to exhibit them.

The immersive era also includes virtual reality (VR) and head-mounted displays (HMDs). Although some use for VR and HMD 3-D has been made in the home environment, these stereoscopic formats have yet to establish themselves as a widespread platform for motion picture narratives.

Digital 3-D Cinema—2005 to Present

It seems evident that the fourth age of stereoscopic cinema began on November 4, 2005, with the release of *Chicken Little 3-D* in eighty-four digital 3-D cinemas, concurrent with its wider release on film in "flat" mode (2-D) to 2,500 theaters.

Stereoscopic movies, along with a few other factors, are driving the proliferation of digital cinemas. As of November 2006, there were 250 digital 3-D cinema screens in the United States; by the summer of 2007, it is expected that 1,000 such screens will be in place.

This book concentrates on the novelty period of stereoscopic cinema. It has been written to demonstrate the rich and varied history of the stereoscopic motion picture before the release of Arch Oboler's 3-D film, *Bwana Devil*, on November 26, 1952. 3-D movies began far in advance of *Bwana Devil*, over a century prior to its release, in fact.

This book also attempts to demonstrate the fundamental importance of stereography to the development of motion picture technology. Stereography preceded photography. The impulse to capture life and replicate it with movement, color, sound, and three dimensions was present at the dawn of photography and motion pictures. This impulse also informs subsequent media from television and computers to the internet.

By examining the uses of stereography in visual history, it is my hope that this book may also provide a critical framework for a stereoscopic grammar of moving pictures. 3-D images present a heightened realism—a visual allure so powerful that they can easily overwhelm the story and subvert the narrative. The novelty period of stereoscopic cinema is a study in the effects of this allure. It was also a century-long laboratory for visionary inventors, utopians of the narrative image, attempting to bring depth to the motion picture screen.

Stereography Begins

Charles Wheatstone—The Binocular Discovery

So often, great scientific breakthroughs seem to be a simple discovery of the obvious. Hidden in plain sight are the mysteries of human perception and stereoscopic vision. The fundamental and powerful fact that we see in 3-D because we have two eyes with binocular vision is just such a discovery. To prove his deduction of this fact, it was necessary for Charles Wheatstone in 1830 to create a device that was to be called the "reflecting mirror stereoscope."

The word "stereoscope" is derived from Greek, conjoining the two words *skopion* and *stereo,* meaning "to see—solid." Wheatstone's stereoscope, utilizing two centered mirrors at 45 degrees to each eye and reflecting right- and left-eye images, was the first instrument designed to view such images and produce a three-dimensional effect.

Binocular vision had been the subject of scientific speculation for centuries. In the third century B.C., Euclid in his treatise on *Optics* observed that the left and right eyes see slightly different views of a sphere.[1] But, as Brian Bowers pointed out in his book *Sir Charles Wheatstone,* "There is nothing, however, to suggest that Euclid understood the stereoscopic effect achieved with binocular vision."[2] In the second century A.D., the physician Galen, with his writing *On the Use of the Different Parts of the Human Body,* noted that a person standing near a column and observing first with the left eye and then with the right eye will see different portions of the background behind the column:

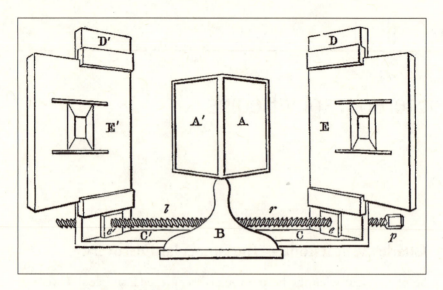

Earliest form of Charles Wheatstone's stereoscope (1833).

> But there are some things seen by the right eye and some by the
> left, and hence the position of the magnitude seen will appear to
> be a property dependent on each of the eyes. . . . Whatever neither
> eye sees will be entirely invisible to both eyes together, and on this
> account the object seen by both eyes at once will conceal less than
> if the eye observing it were alone, whichever eye it is.[3]

Similarly, Leonardo da Vinci (1452–1519) in his *Trattato della Pittura
(Art of Painting)* remarked that a point on a painting plane could never
show relief in the same way as a solid object. "Painters often despair of being able to imitate Nature, from observing, that their pictures have not the
same relief, nor the same life, as natural objects have in a looking-glass,
though they both appear upon a plain surface," wrote da Vinci. "It is impossible that objects in painting should appear with the same relief as those
in the looking-glass, unless we look at them with only one eye."[4]

The first published mention of Wheatstone's stereoscope is in the third
edition of Herbert Mayo's *Outlines of Human Physiology* published in 1833,
which refers to "a paper Mr. Wheatstone is about to publish." It states:

> One of the most remarkable results of Mr. Wheatstone's investiga-
> tions respecting binocular vision is the following. A solid object
> being placed so as to be regarded by both eyes, projects a different

perspective figure on each retina; now if these two perspectives be accurately copied on paper, and presented one to each eye so as to fall on corresponding parts, the original solid figure will be apparently reproduced in such a manner that no effort of the imagination can make it appear as a representation on a plane surface.[5]

Wheatstone first presented his stereoscope to the public before the Royal Society of Great Britain in 1838, where he also presented his historic paper "Contributions to the Physiology of Vision, Part the First: On Some Remarkable, and Hitherto Unobserved, Phenomena of Binocular Vision." With a treatise of 12,000 words, Wheatstone described the stereoscope and claimed as a new fact in his theory of vision the observation that two different pictures are projected on the retinas of the eyes when a single object is seen. He asked, "What would be the visual effect of simultaneously presenting to each eye, instead of the object itself, its projection on a plane surface as it appears to the eye?"[6]

Included with the 1838 paper were a number of line drawings in the form of stereoscopic pairs. Wheatstone had produced these drawings as proof of his theory, and they were made to be viewed stereoscopically in his new invention, which was received very enthusiastically by scientific societies in Britain. Commercial and popular interest in Wheatstone's discovery, however, had to await the invention of photography. In 1852, Wheatstone wrote "Part the Second" to his historic paper. In the meantime, photography itself had been invented.

William Fox Talbot had begun to experiment with fixing photographic images on silver-chloride paper as early as 1835. He called these positive prints "Talbotypes" and later "Calotypes." By 1839, Talbot had announced his discovery to the world, and photographic practices began. As early as 1831, Nicéphore Niepce and Louis Daguerre had begun to fix photographic images on silvered copper plate, and in 1838 these were publicly presented as "Daguerreotypes." In the 1852 paper, Wheatstone wrote: "At the date of the publication of my experiments on binocular vision, the brilliant photographic discoveries of Talbot, Niepce, and Daguerre had not been announced to the world. To illustrate the phenomena of the stereoscope I could therefore, at that time, only employ drawings made by the hands of an artist."[7]

Early in 1839 "the photographic art," as Wheatstone called it, "became known, and soon after, at my request, Mr. Talbot, the inventor and Mr. Collen (one of the first cultivators of the art) obligingly prepared for me

Charles Wheatstone's stereo drawings (1838).

stereoscopic Talbotypes of full-sized statues, buildings, and even portraits of living persons." In 1841, M. Fizeau and M. Claudet of Brussels produced the first daguerreotypes made for the stereoscope. "For obtaining binocular photographic portraits," Wheatstone observed, "it has been found advantageous to employ, simultaneously, two cameras fixed at the proper angular positions."[8] Stereoscopic photography had begun, the offspring of Wheatstone's inquiries into the binocular nature of human vision.

Sir David Brewster—The Retinal Rival

Sir Charles Wheatstone and Sir David Brewster were not the best of friends. During a period of rapid technological change when optical discoveries about 3-D were being made and photography was being invented, these two men were scientific rivals. As Richard L. Gregory writes in his foreword to *Brewster and Wheatstone on Vision,* the two men "did not, as it were by their stereo rivalry, see eye to eye."[9]

Sir David Brewster was a Scottish scientist born in 1781 who is most famous for his invention of the kaleidoscope in 1816, which he explained in his *Treatise on the Kaleidoscope* published that same year. Brewster authored over 300 scientific papers and several books, including his *Treatise on Optics* in 1831 and a definitive biography of Isaac Newton in 1855, and for his efforts he was elected a Fellow of the Royal Society in Great Britain and was knighted in 1831. With his 1856 book, *The Stereoscope: Its History, Theory, and Construction,* Brewster set forth the principles of stereoscopic vision and questioned Wheatstone's claim of priority for the discovery of binocular perception.

Brewster took pains to establish that historical precedents for discoveries similar to Wheatstone's existed and were far from "hitherto unobserved." In particular, he cited the work of Mr. Elliot, a "Teacher of Mathematics" in Edinburgh, who wrote an essay in 1823 that expressed "the idea, that the relief of solid bodies seen by the eye was produced by the union of the dissimilar pictures of them in each eye, but," according to Brewster, "he never imagined that this idea was his own, believing that it was known to every student of vision." Brewster claimed that Elliot had in 1834 conceived of "a simple stereoscope, without lenses or mirrors" consisting of a wooden box 18 inches long, 7 inches wide, and 4 inches high that was "an instrument for uniting two dissimilar pictures." Elliot delayed construction of his stereoscope until 1839 and, since photography had not been invented yet,

Mr. Elliot's stereo pair, as reported by David Brewster (1839).

he "drew the transparency of a landscape"[10] as a stereo pair for viewing in his device. Brewster noted that, in 1852, after perusing Wheatstone's reflecting stereoscope, Elliot "was convinced not only that Wheatstone's theory of the instrument was incorrect, but that his claim to the discovery of the dissimilarity of the images in each eye had no foundation."[11]

Both Wheatstone and Brewster surveyed the writings of Euclid, Galen, and Leonardo da Vinci in their treatises on binocular vision. "Now, although Leonardo da Vinci does not state in so many words that he was aware of the dissimilarity of the two pictures," argued Brewster, "the fact is obvious, [even though da Vinci] was not led by his subject to state the fact at all." Brewster was emphatic in challenging Wheatstone's claim of discovery. "That the dissimilarity of the two pictures is *not a new fact* we have already placed beyond a doubt," he wrote; "and we cannot understand how he failed to observe it in works which he has so often quoted, and in which he professes to have searched for it."[12]

Despite his professional rivalry with Wheatstone, Brewster made significant contributions to stereography. In his book on the stereoscope, Brewster established that the camera lenses in a stereo photograph should have the same aperture as the human eye (about 2/10 of an inch), that the focal lengths of camera and viewer lenses had to be equal, and that the interocular should be about 2½ inches (the average separation between the two human eyes). In 1844, Brewster began making compact stereoscopes with magnifying lenses to fuse stereo daguerreotype pictures taken with a lateral shift of the camera between exposures. By 1849, he had developed

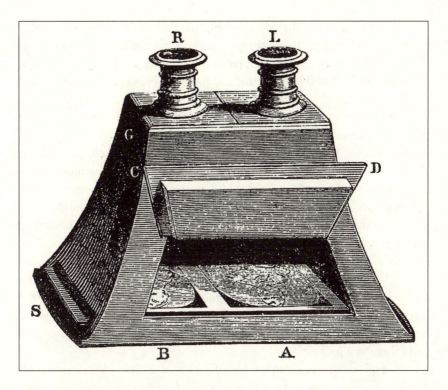

David Brewster's lenticular stereoscope (1851).

his lenticular stereoscope, a box-like instrument with two decentered lenses and a hinged shutter on top to admit light. This stereoscope was first manufactured and sold by Duboscq and Soleil in Paris and was first exhibited in London at the 1851 Exhibition at the Crystal Palace, where it was presented to Queen Victoria, who found it greatly pleasing.

When Brewster's book on the stereoscope was published in 1856, the market for stereoscopic photographs was beginning to boom, and his book included a large listing of stereoscopic pictures offered for sale by the London Stereoscopic Company.

Oliver Wendell Holmes—The Poet of Stereo

"The first effect of looking at a good photograph through the stereoscope," wrote Oliver Wendell Holmes, "is a surprise such as no painting ever produced. The mind feels its way into the very depths of the picture."[13] With an article titled "The Stereoscope and the Stereograph" published in the

June 1859 issue of the *Atlantic Monthly*, this poet of the stereo image cele-
brated the infant art of photography and the mysterious pleasures of the
third dimension. Two years later, he was to produce a simplified version of
the stereoscope that would do much to promote stereo photography to the
world.

One of the most famous American writers of his time, Holmes was also
a teacher, a surgeon, and a lecturer. Holmes's career as a popular writer was
under way in 1857 when James Russell Lowell launched the *Atlantic
Monthly* and assigned the "poet laureate of Boston" to write monthly essays. With his June 1859 article, Holmes was the first to use the word
"stereograph," from the Greek, meaning literally "to write with solids." "We
have now obtained the double-eyed or twin pictures, or STEREOGRAPH, if
we may coin a name," wrote Holmes, and he thought so highly of it that he
characterized it as "the card of introduction to make all mankind acquain-
tances." Holmes wrote in a lyrical fashion about the significance of photography itself: "This is just what the Daguerrotype has done," he said; "it has
fixed the most fleeting of our illusions. . . . The photograph has completed
the triumph, by making a sheet of paper reflect images like a mirror and
hold them as a picture."[14]

In explaining stereography, Holmes created a clear picture of how it
works. "A stereoscope is an instrument which makes surfaces look solid,"
he wrote; "we see something with the second eye which we did not see with
the first." The poet then sketched a beautiful metaphor for the tactility of
stereopsis: "By means of these two different views of an object, the mind,
as it were, feels round it and gets an idea of its solidity. We clasp an object
with our eyes, as with our arms, or with our hands, or with our thumb and
finger, and then we know it to be something more than a surface."[15]

Two years later, Holmes published an article in the *Atlantic Monthly*
titled "Sun-Painting and Sun-Sculpture," which took the reader on a stereoscopic journey around the world and mentioned, in closing, "an instrument which may be held in the hand" that is "very convenient. We have
had one constructed which is better, as we think, than any in the shops."[16]
Holmes's partner in the creation of the inexpensive, open stereoscope was
Joseph L. Bates of 129 Washington Street, Boston, a purveyor of fancy,
fine-quality goods.

Before that time, most stereoscopes were in the form of common
wooden boxes that admitted minimal light for viewing of the stereo image
through a hinged door on the top, as with the Brewster lenticular stereo-

scope. As Holmes wrote in a January 1869 history of the stereoscope for the *Philadelphia Photographer,* his handheld version was "far easier to manage" and "could be made much cheaper than the old fashion contrivances." It included a "hood for the eyes" and was, as Holmes characterized it, a "primeval machine" in its simplicity.[17] Joseph Bates refined Holmes's skeletal scope by adding wires to hold the card on a sliding cardholder for focusing. It was this classic stereoscope that millions used during the golden age of stereography from 1870 to 1920. Holmes lyrically declared: "Form is henceforth divorced from matter. In fact, matter as a visible object is of no great use any longer, except as the mould on which form is shaped. Give us a few negatives of a thing worth seeing, taken from different points of view, and that is all we want of it."[18]

In 1869, Oliver Wendell Holmes had over 1,000 stereo cards in his personal collection, which he greatly valued. He must have had many more than that by 1894 when he passed away, just as the stereoview card was at its cultural zenith. "Oh, infinite volumes of poems that I treasure in this small library of glass and pasteboard!" wrote Holmes; "I creep over the vast features of Rameses. . . . I pass, in a moment, from the banks of the Charles to the ford of the Jordan, and leave my outward frame in the arm-chair at my table, while in spirit I am looking down upon Jerusalem from the Mount of Olives."[19]

The Vortex of Popular Consumption

In 1850, Louis-Desire Blanquart-Evrard had brought to market a paper with a thin, salted coating of albumen that held sharpness of detail in a photographic print made from a negative. It was this paper which was affixed to Holmes's "pasteboard" for production of the standard stereoview.

This albumen paper was used for multiple photographic prints, *cartes des visites,* and for a burgeoning market for stereoview cards, which from 1852 to 1862 became a mass entertainment. Stereograph historian William C. Darrah observed that "by 1862, within a single decade of stereo photography, there were photographers, manufacturers, stereographic emporiums, salesmen, promoters, and collectors. . . . In 1862, the London Stereoscopic Company sold almost a million views."[20]

Photographic historian Reese V. Jenkins writes that "the stereoscopic viewer and box of view cards were as common a feature of the post–Civil War American home as is the television set today." Jenkins also characterizes Wheatstone's original discovery as a "new subjectivist approach to optics."[21] In order for stereography to proliferate as a mass medium, not only did Wheatstone's original unwieldy reflecting mirror stereoscope have to be streamlined, but also a cultural shift in the perception of physiological optics had to take place.

In her essay "From Phantom Image to Perfect Vision: Physiological Optics, Commercial Photography, and the Popularization of the Stereoscope," Laura Burd Schiavo traces the progression of stereography through the invention of photography to popular proliferation as a mass medium:

> Wheatstone had purposefully used eleven pairs of simple line drawn shapes and figures including cubes and cones with his stereoscope, contriving each so that it represented the two projections of an object seen from two points of sight. . . . Because Wheatstone's demonstration sought to display the visual circumstances that contributed to the brain registering dimensionality, he had to isolate the variable of binocular disparity.[22]

Schiavo also contends that when Brewster developed the handheld lenticular stereoscope, he had hidden "the physiological roots of the stereoscopic image" and had transformed the stereoscope into "a commodity form—a popular amusement and a purveyor of unmediated truths."[23] With further refinement and writings by Holmes and others in parlor manuals, advertising materials, and the popular press, the stereoscope came to be seen as a form of "perfect vision," an opportunity to see the world as it is in nature. As a result, in the words of an 1870 issue of *Anthony's Photographic Bulletin,* a "vortex of popular consumption"[24] for stereoviews was set in motion.

By far, the most popular genres of stereograph were portraits and regional scenic views. In a very short time, however, the diversity of subjects became immense. In William Darrah's comprehensive survey, *The World of Stereographs,* an entire chapter is dedicated to "A Subject Guide to Stereographs, Arranged in Encyclopedic Fashion" that includes over seventy-five listings ranging from "Advertising" to "Zoology."[25] Many of the genres, particularly Automobiles, Aviation, Comics, Disasters, Erotica, Exhibitions, Novelties, Theatrical, Railroads, and Wars, were borrowed and would become standard subjects for the nascent motion picture industry during its "novelty period" from 1895 to 1905. "These categories compare quite closely with the general subjects of the motion pictures copyrighted in

1896 by Edison, American Mutoscope and International Film," observes Annette Michelson. She also writes:

> The number of films produced during the first fifteen years of the industry was infinitesimal compared to the vast numbers of stereographs and *cartes des visites* that were produced during the same period. However, the similarity of the important genres that were depicted in the two media is striking, and offers compelling evidence that an imitative relationship existed for some time between the early film producers and the commercial photography industry.[26]

John Fell, in *Film and the Narrative Tradition,* writes: "I believe that in the motion pictures there surfaced an entire tradition of narrative technique which had been developing unsystematically for a hundred years." Fell points out that this narrative tradition appeared in a variety of entertainments but also "in ephemera as diverse as stereograph sets, peep shows, song slides, and postal cards."[27]

The stereograph sets, particularly those produced after 1896 with halftone and lithographic printing, featured a narrative sequence, usually comic, sometimes risqué, of three to twelve images. One-reel films of the motion picture novelty period exploited visual styles and themes of the stereoview cards that had preceded them historically. It was a natural evolution for stereo card genres to migrate into the nascent medium of the motion picture.

Stereoview Peep Shows

After 1870, manufacturers of stereoviewing apparatus began to build cabinets that displayed a sequence of stereoview cards. This sequential viewing of stereoview cards naturally implied a narrative. In his 1975 master's thesis, "Selected Attempts at Stereoscopic Moving Pictures and Their Relationship to the Development of Motion Picture Technology, 1852–1903,"[28] H. Mark Gosser explored the idea that stereoscopy may have been the engine that drove development of motion picture technology. Stereo still peep show cabinets like Alexander Becker's in 1857 (U.S. patent no. 16,962) prefigured Edison's moving picture kinetoscope. Becker's cabinet used stereo cards fastened in series to an endless belt turned by hand, a concept that foreshadows the continuous loop of motion picture film.

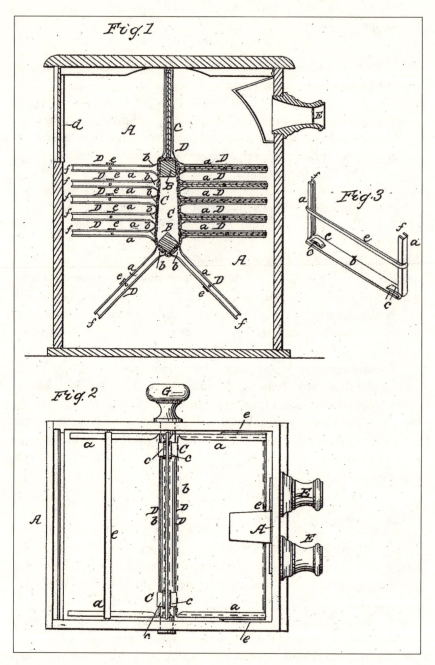

Alexander Becker's stereoview cabinet; knob (G) on the right side advances the stereo cards for individual viewing (1857).

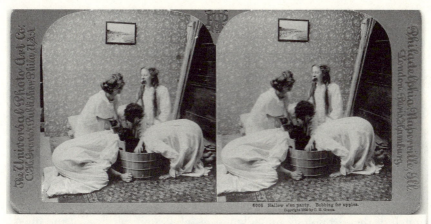

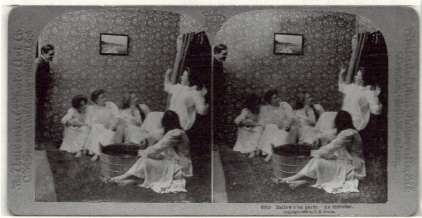

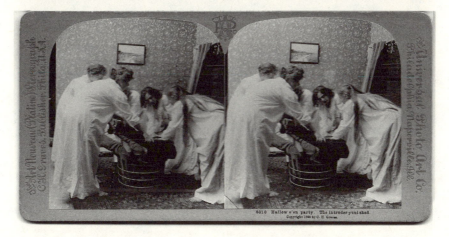

Halloween Party; a narrative in three stereoviews published by C. H. Graves:
bobbing for apples *(top)*, an intruder *(middle)*, and the intruder punished
(bottom) (1899).

The concept of reels or spools of images running between two rollers was also implemented by Frederick Dayton and William S. Kelly in 1862. Their device utilized a wide band of stereoviews unwinding from a top roller within a cabinet, moving past stereoviewing lenses and wound on a take-up roller. Gosser suggests that "from 1852 to 1868 stereoscopy was the key contributing technology to the evolution of a motion picture system."[29]

It is also quite possible that cultural factors drove both the invention of stereography and motion picture technology. The "utopian dream" of capturing "the feeling of life," as characterized by Laurent Mannoni,[30] drove much of the technology underlying the "philosophical toys" of the nineteenth century, which attempted to display moving images with color in three dimensions. This utopian dream also preceded the invention of stereography itself in the form of the non-photographic peep shows that attempted to incorporate both motion and depth.

2

The Peep Show Tradition

An Enclosed Space

Throughout the seventeenth and eighteenth centuries, traveling peep shows were common entertainment to be found at fairgrounds and on the streets in Europe. These changing views within a portable cabinet were an early precursor to the Edison Kinetoscope and motion pictures. Not surprisingly, seventeenth-century inventors and showmen attempted to combine the perception of depth with the illusion of motion in their novel entertainments.

Historian Richard Balzer describes a seventeenth-century peep show created by Carel Fabritius and Samuel van Hoogstraten that is on view at the National Gallery in London:

> The box is open on one side allowing the entrance of light and a non-distorted view of the interior. Better is the view from the peep holes in two opposite corners of the box. The room comes to life. The dog, ingeniously painted on the floor and up one side of the wall, sits up; a chair, similarly painted, appears three dimensional; flat space seems curved; paintings appear on walls seemingly from nowhere; and a roof painted on two surfaces takes on a three dimensional appearance. One is treated to two quite different vantage points. The mix of optical illusion and artistic talent combines to create a very powerful impact.[1]

Many of the peep boxes had used perspective and mirrors to convey depth so that a tradition of spatial illusion was inaugurated. "With them, too, the

moving picture was for the first time enclosed in a definite frame," notes
Olive Cook; "and peepshows and panoramas were not projections of three-
dimensional bodies, they were essentially two-dimensional images to which
a third dimension was added by means of perspective, special lenses, and
transmitted or reflected light."[2] Like the peep shows that had preceded it,
the stereoscope provided the experience of viewing into an enclosed space.
The experience of looking into a private space, as with a keyhole, shut out
everything but the enclosed image. This visual exclusion of the world might
have provided a sudden notion of revelation, as if exposing something that
had been hidden.

The poet Charles Baudelaire, no friend of the naturalistic school of art,
denounced the stereoscope in 1859, by which time erotic stereo daguerreotypes and stereoview cards had begun to be published in France. "It was not
long before thousands of pairs of greedy eyes were glued to the peep-holes
of the stereoscope, as though they were the skylights of the infinite," fulmi-
nated the poet; "the love of obscenity, which is as vigorous in the heart of
natural man as self-love, could not let slip such a glorious satisfaction."[3]

In 1822, Louis Daguerre had invented a very successful optical entertainment called the Diorama. It was exhibited in an auditorium and used a
wall-sized transparent painting visible through a single proscenium under
changing light, which revolved with the spectators to slowly move from
one part of the painting to another. By 1860, the Stereorama, an optical
device based on the Diorama, was in use. With the Stereorama, sequences
of stereoview cards were exhibited and viewed in rotation, successively, by
four spectators at the same time.

The French tissue stereoview cards produced in the last half of the
nineteenth century in Paris featured hidden magic transformations in this
manner. The major producer was BK, whose editor was identified as "A.
Block." BK released sets of six or twelve tissue stereoview cards in a series
called "Le Theatres de Paris." These theatrical stereoview tissues featured
narratives such as Cinderella, William Tell, and Faust. The 3 ½ by 7–inch
cards themselves had a die-cut open back and a stereo photograph of elaborate tableaus created with miniature sculptures. The stereo photo was
printed on thin paper backed by hand-colored tissue that was also frequently pierced with a pin. With front-light viewing, the scene was black
and white. When held up to a backlight, the scene magically transformed
to color with light effects such as glowing eyes or stars from the piercing. In
addition to their use of popular narratives, the visual effects in the French

Front *(above)* and back *(below)* of a nineteenth-century French tissue stereo-view; the tinting and piercing are visible on the open back.

tissue stereoviews foreshadowed many of the techniques of cinema, such as the "dissolve" and time-lapse photography that depicted the passage of day into night.

The Stereoscope Cosmorama Exhibit opened at 188 Chestnut Street in Philadelphia before the Civil War. "There each spectator sat," notes Martin Quigley Jr., "and could see one stereoscopic view after another by turning a crank."[4] The Stereoscope Cosmorama was produced by Frederic and William Langenheim, two brothers who did much to popularize the stereoscope and published many stereoview cards. Several of these rare optical devices and toys were once housed in the Barnes Museum of Cinematography in Great Britain.[5]

Around 1860 in America, D. Appleton and Co. of New York manufactured and sold a three-part cardboard assembly stereoscope with a series of lithograph views it called "The Stereoscope for the Millions." On the base of the stereoscope was printed an advertisement for "The Motoscope, with 12 Views: Exhibiting the Figures in the Pictures in Motion" selling for 62 cents. The Motoscope advertisement included "Directions for Its Use," instructing the potential purchaser to "Push the wire at the right of the Stereoscope back and forth at the speed of about once every two seconds when *the figures will be seen in motion.*"[6]

On November 22, 2000, stereoscopic historian Paul Wing sent me a photo of the lens board to the Motoscope, along with a lithographic view card for use in it. "A picture of a well worn lens board shows a spring loaded push button at the upper right which gets rid of the need for rapid blinking," wrote Wing; "the viewer is in three parts—the lens board—a simple flat base—and a hinged card holder at the rear." Wing notes that a series of stereographs were produced in France, with motion depicted between the left- and right-eye images. These rare early views were reproduced by lithography and then sold by D. Appleton and Co. for use with the Motoscope. "The copy I have mounted and sent along shows two marks on the bottom where holes were punched for mounting in set using thin ribbon."[7] The stereoview card depicts a violinist in two different positions and bears the title "An Enthusiastic Musician." The addition of motion to these enclosed stereoscopic spaces could only enhance the sense of life, what may have seemed a privileged and intimate glimpse of nature.

Persistence of Vision

The fundamental principle behind the movies is persistence of vision, when a visual impression remains briefly in the brain after it has been withdrawn. This principle was demonstrated in 1825 with an optical toy called the "Thaumatrope," invented by Dr. John Ayrton Paris. The Thaumatrope consisted of a paper disk with different images printed on opposite sides. Strings attached on either side of the disk enabled it to be spun so that persistence of vision made the two different images merge into a single picture. A rider on one side of the disk would sit atop a horse printed on the opposite side, or a bird would appear in a cage.

In 1890, Professor David Wells of the Massachusetts Institute of Technology created stereoscopic charts to aid students in comprehending solid

Motoscope viewer: an enthusiastic musician in a two-step motion stereoview card *(above);* lens board for the viewer—the pushbutton on the right is used to open and close the left and right eye alternately *(below)* (1860).

geometry. These stereoscopic charts were eventually compiled into an oph-thalmological set to be viewed in the Holmes Stereoscope for binocular vision training. Wells used Paris's imagery from the Thaumatrope—the bird in the cage, a horse and rider—for several of his stereoscopic charts, which were first published in 1912 by the American Optical Company.[8]

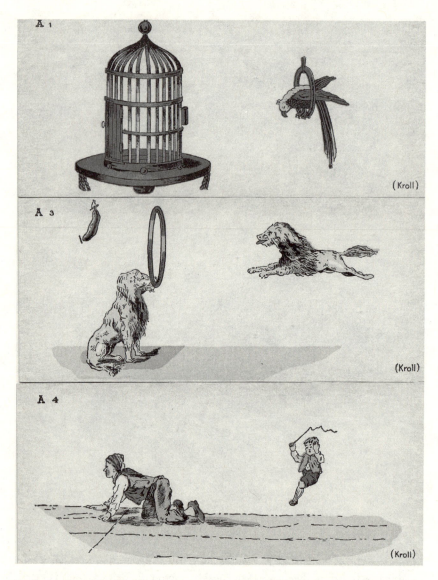

Professor David Wells's stereo test cards (1912) borrowed imagery from Dr. John Ayrton Paris's Thaumatrope (1825).

The "Wheel of Life" or "Zoetrope" developed by William Horner in England in 1834 used the principle of persistence of vision with a series of slits cut in the sides of a spinning cylinder through which the viewer perceived sequential images drawn on strips of paper. Rotating cylindrical drums and disks were later to be used by different inventors exploiting persistence of vision in devices designed to display stereoscopic moving pictures. The Phenakistoscope, invented by Joseph Plateau in 1831, was a predecessor of the Zoetrope, using slits cut on a rotating disk viewed before a mirror. At about the same time, a Viennese mathematician, Simon von Stampfer, created a similar device for an optical toy he called the "Stroboscope."

In 1849, Plateau reported that Wheatstone had contacted him with an idea "which consists of a combination of the principle of the stereoscope with that of the phenakistoscope." Shapes "painted on paper would inevitably be seen in three dimensions and moving [and] would thus entirely present all the appearances of life. It would be the illusion of art taken to its highest point." In addition, Wheatstone proposed to Plateau that sixteen stereo daguerreotypes of a plaster figure that was incrementally moved could also be used for the device. "This would undoubtedly require a long effort," wrote Plateau, "but it would be rewarded by the marvelous nature of the results."[9] No evidence exists, however, that Plateau followed through with development of a stereo Phenakistoscope.

Wheatstone was not aware of the efforts of Antoine Claudet in the early 1850s to build a stereoscopic motion picture device. In the letters column of an 1852 issue of *La Lumiere,* correspondence is noted from Claudet on his attempts to create such a device:

> M. Claudet writes us that he has constructed a stereoscope in which one sees moving figures, for example a woman working with a needle and making all the necessary movements, a smoker moving his cigar in and out of his mouth, during which he exhales the smoke and which he blows outside, people who drink and offer toasts in the English manner, steam engines in motion, etc. etc."[10]

While Claudet was living in London in 1852 he unveiled what he called his "fantascopic stereoscope," which could "obtain three dimensions with movement."[11] On March 23, 1853, Claudet secured British patent no. 711 for the device titled "Improvements for the Stereoscope," which described a Brewster-type stereoscope in which stereoviews turning on a central shaft and fixed to cross-shaped forms were sequentially revealed with a moving bar. Eyeholes in the bar alternately blocked or revealed the images.

In a highly influential 1865 article, "Moving Photographic Figures," that was reprinted many times in scientific journals and trade publications, Claudet wrote about his fantascopic stereoscope as follows:

> But nothing is so easy as to employ eight different pictures in as many different stages of action, and with this number of pictures the effect will be sufficiently complete. For this, having placed in the stereoscope two separate cubic frames, revolving independently on the same horizontal axis, I have only to fix on their four sides at right angles two sets of four pictures, making eight pictures, which are made to pass in consecutive order, one after the other, before the lenses of the instrument, and the figure will appear to assume consecutively eight different stages of the whole action.[12]

Claudet was not the first to patent such an invention. That honor belonged to his fellow countryman, the Parisian optician Jules Duboscq. Claudet's fantascopic stereoscope did not become a success. By the time his article appeared, over a decade later, there had been numerous developments with stereoscopic motion picture devices.

The First Stereoscopic Motion Picture Patent

In partnership with Francois Moigno, Jules Duboscq had successfully marketed Brewster's refracting stereoscope in 1851. With an addition of November 12, 1852, to his French patent no. 13,069 dated February 16, 1852, Duboscq described the "stereoscope-fantascope or Bioscope," a device that combined "the essential properties of the stereoscope with the most wonderful properties of M. Plateau's Phenakistoscope." In an article of 1857, Duboscq wrote that the instrument created "the feeling of three dimensions and movement, or the feeling of life."[13]

The Bioscope used a rotating disk with twelve stereo pairs arranged around the circumference. The left- and right-eye images were vertically stacked for viewing through mirrors that reflected the images in the proper orientation. In his article of 1865, Claudet described the operation of the Bioscope:

He [Duboscq] had fixed two series of binocular photographs on two zones of the revolving disk of the phenakistoscope one above the other, and by means of two small mirrors, placed each respectively at the inclination capable of reflecting the two zones, on the same horizontal line, from whence the images could separately meet the axes of each of the two prismatic lenses of the stereoscope.[14]

A photo of the only surviving Bioscope disk is in the book *Paris in 3D,*[15] and no record of the apparatus or method used to produce the stereo photos remains. The prints themselves are albumen paper; the black pasteboard disk, in the possession of the Plateau Collection at the University of Ghent, Belgium, measures 33.5 centimeters in diameter.

The Bioscope was not a commercial success. Duboscq himself commented about this fact in his article of 1857: "There is no reason to hope that the bioscope, like the panoramic stereoscope, could ever be sold in very great numbers," he wrote, "since they require a considerable number of proofs, as many as 24 or 32 to obtain a single effect."[16]

Duboscq also constructed a second version of the Bioscope that used a rotating cyclinder in the style of the Zoetrope. Claudet also described this apparatus in his article of 1865:

> Instead of the vertical original revolving disk of Plateau, he [Duboscq] employed a cylinder revolving on its vertical axis, and he placed on two zones of that cylinder, one above the other, the two series of photographic pictures between the slits through which the eyes can see the pictures, and by means of two mirrors, as in the other apparatus, each series was reflected on its respective lens through the cylinder.[17]

Gosser points out that both Duboscq's machines were the first patented devices to use photography for the display of motion pictures. "It is significant," he observes, "that Duboscq chose stereoscopic photos instead of conventional flat photography."[18]

Disks and Cylinders

In England, William Thomas Shaw in May 1860 patented two forms of 3-D peep shows, one of which used the Phenakistoscope application.

Shaw's first apparatus "featured one member of the stereo pair mounted on each of two disks inside a cabinet. The images were viewed with two mirrors mounted at 45 degrees to each eye, as in Wheatstone's stereoscope."[19] Shaw also built a design that used the system of Sir David Brewster's lenticular stereoscope with eight stereographs of successive motion mounted inside an octagonal drum. Because Shaw made no provision for intermittent motion with his device, it was not a great success.

The utopian dream of stereoscopic images in motion was very much alive from 1850 to 1875, and numerous inventors and photographic specialists worked on systems to achieve it. A Bohemian inventor named Johann Czermak created a form of vertical Zoetrope in 1855, which he called the "stereophoroskop," that used an octagonal wheel with slits and internal mirrors that reflected stereo pairs of images. In 1853, Parisian painter Philippe Benoist built an "animated" stereoscope.[20] Adam Jundzill, an engineer in London, with his "Kinimoscope" device attempted to animate stereoscopic cartoons in 1856.[21] With a French patent of 1859 for an "Animated Stereoscope," Victor Pierre Sies, a toymaker, created a device for hand-colored albumen prints in which "four apparently similar photographic illustrations falling and rising can create the illusion of movement."[22] The Paris company of Fume and Tournier produced a stereoscopic Phenakistoscope.[23] Andre David, also in Paris, patented a device in 1873 based on Claudet's process.[24]

Belgian civil engineer Henry DuMont was among the most ambitious of the utopian inventors. In 1859, DuMont secured patents in France and Belgium for nine different designs of stereoscopic motion picture devices.[25] "The basic forms were discs and cylinders," writes Gosser, "though DuMont used the outside surface of his cylinders rather than the inside wall of the usual Zoetrope. In most cases, a slotted revolving shutter disk was used to reveal the images (in some cases transparent images) and freeze them for that split second when the image formed on the retina."[26] DuMont gave the name "Omniscope" to his several devices. He was a proficient optical engineer. One design used a twin, synchronized Zoetrope combined with Wheatstone mirrors in the middle at a 45-degree angle. Another design used twin Zoetropes and an internal projection system with a mirror for viewing with red-green anaglyph glasses.

In 1861, DuMont obtained three additional patents that described a system with a band of stereoscopic photographs that moved intermittently. In an 1862 article, DuMont remarked on the need for recording the stereo photographs at a faster rate and "a new photographic apparatus, which permits the reproduction of successive phases of a movement at only several fractions of a second intervals."[27]

By 1867, Wheatstone's concept of a stereo Phenakistoscope was realized by Henry Cook and Gaetano Bonelli with a device they called the "photobioscope" that used translucent glass disks with two rows of sequential photographs of a sculpture being rotated. Produced with wet collodion photography, one of these disks survives today, along with a stereoscope used for viewing them, at the Musee National des Arts et Metiers in Paris. Cook exhibited the photobioscope at the Societe Francaise de Photographie on August 2, 1867.

Using a form of Zoetrope with printed bands of figures lining the inside of a rotating drum with mirrors set at angles in the center to reflect moving images, Emile Reynaud created in 1877 what he called the "Praxinoscope Theater." By 1889, Reynaud was projecting these images for audiences with narrative entertainments in his "Theater Optique." In using an endless band of pictures, Reynaud's Theater Optique foreshadowed the technology of the motion picture. Also like his peep show forebears, Reynaud was interested in stereoscopic moving pictures. To display them, he reverted to phenakistoscopic technology using two drums, each carrying a separate-eye view of the movement. By 1907, Reynaud had built his "Stereoscopic Binocular Praxinoscope," a device with a mirror complex mounted between two pie plate–shaped cylinders and a binocular eyepiece mounted to the front for viewing of the stereo pairs.

Reynaud built a stereoscopic motion picture camera with which he filmed himself, a 3-D self-portrait in motion, specifically for viewing in the Stereoscopic Praxinoscope. The inventor became despondent over financial difficulties after its invention, however, and threw the camera and many of his other inventions into the Seine River. Only one stereo Praxinoscope was ever built and, fortunately, it survives today. It is housed, along with the strips of stereoscopic drawings and photographs, at the Musee des Artes et Metiers in France.

In his 1865 article, with great foresight, Claudet had delimited some of the inherent technical problems for stereoscopic motion pictures that used disk or cylinder display technology. With the phenakistoscopic disk, the left- and right-eye pictures, because they were vertically stacked, did not have equal radii and did not rotate at quite the same speed. With side-by-side stereo photos on the inside of a curved surface inside a cylindrical drum, the focus is always slightly different between the left- and right-eye pictures.

These limitations could be dealt with optically, of course, but an immediate answer to the problem could be shown in the form of a stereo peep box that would incorporate intermittent motion with an endless band of continuous images. The stereo still peep shows that had proliferated in the 1850s exerted an influence on this development.

Endless Bands

From 1849 to 1852, Charles Wheatstone had experimented with a stereo Phenakistoscope. Quigley claims that "Wheatstone also developed a combination of photos and the Plateau disk which was fitted with a cog which made each photo rest momentarily as it was held before the mirror. The same instrument was made in France under the name of Heliocinegraphe." Unfortunately, this device, if it did exist, has been destroyed. If it was actually built, it would be the earliest example of intermittent motion on record. Acknowledging that Wheatstone "had a marked influence on magic picture development during the middle part of the 19th century," Quigley also suggests that "it may well be that the efforts expended in trying to combine the third dimensional effect of his stereoscope with the magic disk retarded development of screen projection of motion pictures."[28]

In 1870, Wheatstone did eventually build a stereo peep show similar to Shaw's. According to Gosser, "in Wheatstone's apparatus, pictures were placed on an endless band which was fastened to the outside of a spoked wheel. The viewer watched the stereo photographs through a twin eyepiece at the top front of a wooden peep-show cabinet."[29] To create the intermittent motion, Wheatstone used a toothed wheel and a pawl-like stop device. There was no shutter or period of rest for the image in front of the eyes, so the motion effect was less than desirable. Only one of the devices was ever built, and it was never patented. At one time, along with a strip of stereo pairs, it was in the possession of the Science Museum in London.

With a British patent (no. 537) dated February 27, 1860, Peter Hubert Desvignes was the first inventor to describe a stereoscopic motion picture device that used endless bands. In a patent that included twenty-eight different designs for monocular and stereoscopic Zoetrope devices, the last figure depicted an endless band running between two spools. There is a good possibility that Desvignes was adapting the endless band idea from still stereoscopic peep boxes for motion pictures. In addition, Desvignes suggested the use of an electric spark to periodically light the endless band of moving images and thereby create intermittency. This concept may not have been "reduced to practice" by Desvignes, but it foreshadows the 1893 "electrical tachyscope" of Otto Anschutz, who used a Geissler tube to create intermittency with a flashing light.

In a footnote, Gosser remarks that "one must also consider the work of Pierre Czugajewicz." Gosser, noting an article in *Photographic News,* recounts that the journalist dubbed Czugajewicz's device a "Russian Stereoscope" and that "it, apparently, used two endless bands with one member of the stereo pair on each band. There was also a proposal for a camera which utilized 'rolling bands.'"[30]

In the United States, engineer Coleman Sellers patented in 1861 a stereoscopic moving picture peep show that he called the "Kinematoscope." Sellers, like the Langenheims, resided in the city of Philadelphia, home of their "Stereoscope Cosmorama Exhibit." Quigley surmises that "it may very well have been this turning crank system which suggested an interesting motion picture device to the fellow citizen of Langenheims, Coleman Sellers."[31]

Sellers discovered the important principle of intermittent motion for moving pictures while building this stereo Zoetrope and wrote, " it is absolutely necessary, that the pictures should be entirely at rest during the moment of vision, that is advancing toward the eye, or receding from it." With Sellers's 3-D peep show, the stereographs were arranged within a cylinder cut away on two sides to admit light. Stereographs were placed on a series of wings radiating from a center spindle or wheel and viewed through slits in succession "as they advance toward or recede from the eye in the direction of the line of vision." Sellers noted that "this form of instrument has a great advantage in keeping the picture in view for a long time for the picture cannot be seen entire through these slits except when the wheel is in motion."[32]

It was Sellers's goal to depict stereoscopic motion photographs of his children at play. To achieve this, it was necessary to make still stereo photographs showing successive phases of action. Sellers managed to capture his two sons at play, one rocking in a chair and the other pounding with a hammer. He arranged the stereo photographs in proper order to convey the most realistic motion effect. Once he had completed this goal, the inventor lost interest in the optical toy he had created. Outbreak of the Civil War in America undoubtedly limited marketing efforts for Sellers's kinematoscope.

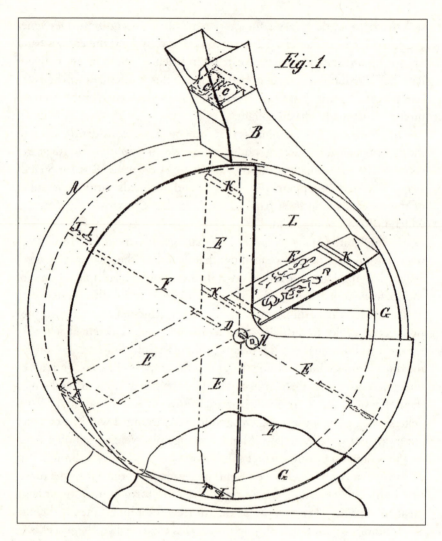

Coleman Sellers's stereo movie peep show (1861).

With an email of March 2005, film historian Stephen Herbert of Great Britain provided an account of a March 1864 meeting of the Royal Scottish Society of Arts in which a communication from James Laing, a mechanic from Dundee, Scotland, was read in conjunction with a demonstration of his "Motoroscope" instrument.[33] In constructing the Motoroscope, Laing's intention "was to give motion, besides relief, to the individual objects of stereoscopic pictures." Laing noted that "the motoroscope is a combination of two previously known instruments, viz., M. Plateau's Phenakistoscope, or magic disc, and the stereoscope." Interestingly, the inventor had independently discovered the inherent defects of both the disk and cylinder displays for moving stereographic images. His solution was to dispense with the use of a cylinder and to "pass a revolving web over two rollers of such a diameter as suits the focal distance of the stereoscopic eyeglasses." Laing discovered that the breadth of the slits in the revolving web was critical, and he constructed slits, both in the revolving web and the eyepieces, that were sufficiently narrow to create great "distinctness" in the stereoscopic pictures.[34]

"As to the producing of a sufficient number of stereoscopic pictures in order to make the instrument an amusing one for in-door pastime," wrote Laing, "there is certainly a difficulty." Laing's solution was to create stop-motion animation with wooden models and "smoke from the cottage chimney being produced by winding a quantity of white wool round a wire, and bending it into required positions." A flag made of paper and mill fans of wood were photographed stereoscopically with "various movements being made at each successive impression."[35]

Laing's Motoroscope demonstration of "stereoscopic pictures in motion" at the meeting "excited considerable interest."[36]

The Utopian Dream

In the later years of the nineteenth century, the work of many showmen and inventors was combined to produce the modern entertainment of the twentieth century. For these pioneers, as for audiences of the twenty-first century, stereographic moving images represented an alluring pinnacle in visual entertainment.

The period from 1852 to 1875, particularly in France, saw great activity, with the invention of numerous stereoscopic optical toys designed to achieve a synthesis of motion, color, and the third dimension. In the absence of a means to create stereoscopic motion photography, the images were drawn and hand colored.

The stereo peep show movie devices established the need for an endless band or revolving web technology that prefigured the invention of a continuously moving strip of film, the technological standard that would make possible the establishment of the motion picture industry itself.

A utopian sense of a new world of visual reality informed the many efforts of inventors. It was a widespread and general attempt to capture the image of life itself. This widely prevalent sense is suggested by the names of the many stereoscopic movie devices: Fantascopic Stereoscope, Bioscope, Motoscope, Diorama, Cosmorama, to name a few.

The peep show tradition continues in the twenty-first century world of stereography with virtual reality and the head-mounted display, both of which provide immersion in a stereoscopic world for the user. They are highly complex technological successors to the stereoscope of the nineteenth century.

3

Motion Pictures Begin

Eadweard Muybridge—Capturing Time

When Coleman Sellers arranged the successive still stereo photographs of his sons at play in the nursery, he attempted to place them in the proper order to convey motion. "A little experimenting showed that a better illusion could be secured if three photos were made of the hammer at the beginning of the stroke and two at the middle, with one illustrating the hammer achieving its purpose," writes Homer Croy; "arranged in this order they better conveyed the increasing swiftness of the stroke." Significantly, Croy characterizes the Sellers stereo photographs as "the first photo ever made to show motion."[1]

Eadweard Muybridge, also known as "Helios," formerly Edward Muggeridge of Kingston-on-Thames, England, was faced with a similar challenge in his attempt to photograph successive images of a horse in motion. The story, practically a legend in motion picture history, asserts that Muybridge, backed by Southern Pacific railroad tycoon Leland Stanford, was led to set up a series of instantaneous photographs in 1872 to settle a bet between horsemen as to whether a horse's four legs are ever off the ground at once.

Muybridge has been a controversial figure in motion picture histories. He certainly didn't fare too well in Terry Ramsaye's *A Million and One Nights,* a 1926 two-volume history of the early motion picture. "Muybridge, in a word, had nothing to do with the motion picture at all," wrote Ramsaye, "and, in truth, but a very small part, if any, in the creative work of the hallowed race horse incident."[2] "Many strenuous attempts have been made by American writers to decry Muybridge's work," wrote Leslie Wood;

"the true importance of the pictures of the galloping horse lies in the fact that they were photographs of real and continuous movement and not posed pictures to counterfeit action."[3]

Ramsaye's account was very much influenced by Thomas Edison, who possibly was looking for a way to disavow the fact that Muybridge had visited him at his laboratories in Orange, New Jersey, on February 27, 1888. On that date, Muybridge had given Edison a demonstration of his animal locomotion studies using his projecting Zoopraxiscope. "Edison's meeting with Muybridge was important enough to the inventor," writes Rebecca Solnit, "that he afterward denied what happened and tried to shuffle the sequence of events."[4] Solnit, most recently, and Gordon Hendricks,[5] a strong critic of Thomas Edison, have published books that have defended Muybridge's work and championed its importance to the evolution of the motion picture.

After arriving in San Francisco in 1855 and spending over a decade as a bookseller, Muybridge took up stereo photography in 1867. Using the *nom de camera* "Helios," Muybridge advertised that he was "prepared to photograph private residences, animals, or views in the city or any part of the coast."[6] He shot both 6 by 6 large-format and stereo photographs in sequence. One of his first projects, following in the tracks of stereo photographer Carleton Watkins, was a series of 160 stereoviews of Yosemite National Park.

The stereoviews of Helios were sold on Montgomery Street in San Francisco, where there was a growing market for stereo photographs. "The majority of the landscapes taken by Muybridge and many of his peers would be in this hyperdimensional medium," writes Solnit: "The effect was less like the depth of field and dimensionality of ordinary binocular vision than it was like the pop-out valentines and paper theaters popular in the era. . . . Stereocards fed a passionate desire to see the world represented as compelling as possible, a desire that would find its greatest satisfaction in movies."[7]

By 1869, Muybridge was experimenting with a spring-operated shutter that allowed for fast exposures of clouds, of which he made a series of fifteen studies in stereo. Solnit points out that Muybridge was interested in capturing ephemeral phenomena, in nature as well as in the city. "It is easy to see precursors of the motion studies in this early work," she writes; "he was already preoccupied with how photography could capture time—not a single moment of time already past, as a single photograph does, but the transformations wrought by time's passage."[8]

Stereoview card by Eadweard Muybridge (ca. 1860).

Using a series of horizontally spaced cameras with shutters tripped by wires, Muybridge began to make photographic motion studies of a horse trotting in 1872. He was subsequently hired by Leland Stanford at Palo Alto, California, based on the recommendation of newspaper editor Fred MacCrellish and a rising reputation as a photographer of the West. Initially working with a wet plate collodion process, it wasn't until 1877 that Muybridge was able to accelerate the camera exposure sufficiently to photograph animal motion in sequence.

By 1877, Muybridge had also produced a series of stereo panoramas of San Francisco and Guatemala, forerunners to the cinematic experience. "The stereocard panoramas," writes Solnit, "suggest that the viewer would . . . , while keeping the stereoscope clapped to his or her eyes, change the cards in sequence to create what cinematographers call a 'pan' of a place, and some stereoviewers were designed to feed a sequence of cards into the viewer."[9]

Did Muybridge ever produce any motion studies in stereo? Gosser notes that Muybridge "apparently tried a battery of stereo cameras around 1877–1878" while working for Stanford at Palo Alto and cites the preface to Muybridge's 1899 book *Animals in Motion,* where the photographer wrote, "Each of the cameras used at this time had two lenses, and made stereoscopic pictures." "It is logical to assume," concludes Gosser, "that he [Muybridge] shot and viewed stereo moving pictures at some point during his work at Palo Alto. He thus became the first researcher to combine real photographed motion and relief."[10]

During the final years of his life, Muybridge returned to his house at Kingston-on-Thames with the remains of his work. In 1998, fragments of glass plate negatives for a stereoscopic motion study of a man riding with horse and buggy were unearthed from the garden behind the house. One of these negatives is depicted in the book *Eadweard Muybridge: The Kingston Museum Bequest,* edited by Stephen Herbert.[11] This negative is nearly identical to one of six cards of Stanford's horses in motion that Morse's Gallery in San Francisco published (and sold for $2.50 each or the set of six for $15) in 1878 with the title "Abe Edgington," undoubtedly the name of the driver of the sulky.

Etienne-Jules Marey—Chronophotographer

When Muybridge's motion studies were published in the December 28, 1878, issue of *La Nature,* Etienne-Jules Marey wrote to the magazine requesting permission to get in touch with their author. "I would like his assistance," wrote Marey, "in the solution of certain problems of physiology too difficult to resolve by other methods." Marey suggested the use of the Zoetrope to view Muybridge's motion studies. With the device, "one could see all imaginable animals during their true movements," he speculated, and "it would be animated zoology."[12]

By 1880, Muybridge had created a device combining photography, the Zoetrope, and the magic lantern, which he called initially the "Zoogyroscope" and then the "Zoopraxinoscope," to publicly project his moving images. The Zoopraxinoscope used a strong projector with a 16-inch disk carrying sequential images rotating in one direction and a second, slotted disk revolving in the opposite direction to create the required intermittency for persistence of vision. No stereoscopic exhibition with the Zoopraxinoscope, however, is known to have taken place.

Marey was primarily interested in human and animal movement, and he invented a chronophotographic gun that recorded rapid exposures on a rotating glass plate to capture phases of motion while tracking the movements of birds in flight. Marey also found stereoscopic photography to be a useful tool for the study of human and animal movement. As early as 1870, Marey worked with his student Gaston Carlet to create stereoscopic pairs of line art to analyze and synthesize images of human motion. One instance illustrates a man's pubic bone and its movements in a trajectory toward the stereo camera. In 1885, using a shutter disk inside a stereoscopic lens board, Marey recorded figures of a running and walking man. "Seen in the stereoscope," wrote Marey, the figures "are perfectly three dimensional: it is as though one is looking at a wire twisted in various directions and periodically repeating the same inflections."[13]

To illustrate the creation of geometrical figures from human movement, Marey used a stereoscopic chamber and camera shutters to create an extended "timelapse" exposure of a white wire moving over a black background. "Ultimately, it is probably Marey who best symbolizes the meeting of stereoscopy and photographic animation in the nineteenth century," writes Mannoni; "his 1887 zootrope, in which he showed eleven statuettes made using chronophotographs representing a seagull in eleven stages of flight, offers an unforgettable sight: the bird is seen with amazing clarity, in three dimensions, flying in slow motion."[14]

In his 1895 book *Le Mouvement,* Marey discusses both still stereo photography and chronophotography for analysis of the "stereoscopic trajectory" of birds in flight and points out that two-dimensional methods are not fully illustrative of movement. With this work, Marey was proposing the efficacy of stereo-chronophotography for the study of motion.[15]

Edison's Kinetoscope

By November 1877, Thomas Edison had perfected his phonograph. As early as December 2, 1877, in the pages of *Scientific American* magazine, an editor made a proposal that reflected what William Kennedy-Laurie Dickson would later characterize as the "crowning point of realism" for the nascent audiovisual media of the late nineteenth century. "It is already possible," wrote the editor, "by ingenious optical contrivances, to throw stereoscopic photographs of people on screens in full view of an audience. Add the talking phonograph to counterfeit their voices and it would be difficult to carry the illusion of real presence much further."[16]

With his first Motion Picture Caveat[17] dated October 8, 1888, and filed with the U.S. Patent Office on October 17, Edison wrote, "I am experimenting upon an instrument which does for the Eye what the phonograph does for the Ear, which is the recording and reproducing of things in motion, and in such a form as to be both Cheap practical and convenient. This apparatus I call a Kinetoscope 'Moving View.'"[18] In this caveat, Edison was considering recording images on a disk or cylinder as with the phonograph. By the time Edison filed Motion Picture Caveat IV, dated November 2,

1889, he had returned from Europe where he visited Marey who was experimenting at the time with strip paper and film for chronophotography. In Caveat IV, Edison wrote about the use of transparent "sensitive film" that was "in the form of a long band passing from one reel to another in front of a square slit."[19]

William Kennedy-Laurie Dickson came to work for Edison in West Orange, New Jersey, in 1883. As one of Thomas Edison's assistants at the New Jersey laboratories, Dickson was assigned by Edison to concentrate on developing a "kinetograph" to photograph motion in 1889. Dickson and the "Boss" were the only two people with access to Room Five where the kinetograph development took place. In an 1895 history of the Kineto-scope, William K.-L. and Antonia Dickson wrote, "The crowning point of realism was attained on the occasion of Mr. Edison's return from the Paris Exposition of 1889, when Mr. Dickson himself stepped out on the screen, raised his hat and smiled, while uttering the words of greeting, 'Good morning, Mr. Edison, glad to see you back. I hope you are satisfied with the kineto-phonograph.'"[20] The authors explicitly described a projected motion picture on the screen synchronized with sound from a phonograph record. They discussed further the new miracle of 3-D motion pictures on the screen: "Projected stereoscopically the results are even more realistic, as those acquainted with that class of phenomena may imagine, and a pleasing rotundity is apparent which in ordinary photographic displays is conspicuous by its absence."[21]

In reality, Dickson and Edison had yet to achieve viable screen projection of motion pictures. Edison himself did not for years afterward believe that motion pictures should be projected on the screen. It was Dickson who, while Edison was in Europe, had built the "Black Maria," the world's first motion picture studio at Edison's laboratory. Edison instructed Dickson to concentrate on a "peep show" machine, a box enclosing rollers, strips of film, and a shutter, which eventually became the Kinetoscope. A letter of reply written June 16, 1891, to a New Jersey man from Edison labs attested: "It is Mr. Edison's intention to give a stereoscopic effect to the pictures taken in connection with the Kinetograph, and a long extensive series of experiments have been conducted at the Laboratory, very good results being obtained. This has all been incorporated in the patent."[22]

In the late 1880s, Dickson had kept a notebook, which Hendricks identified as the "Dickson Kinetoscope Notebook." In this notebook, on page 47, the first entries about the Kinetoscope appear with the following

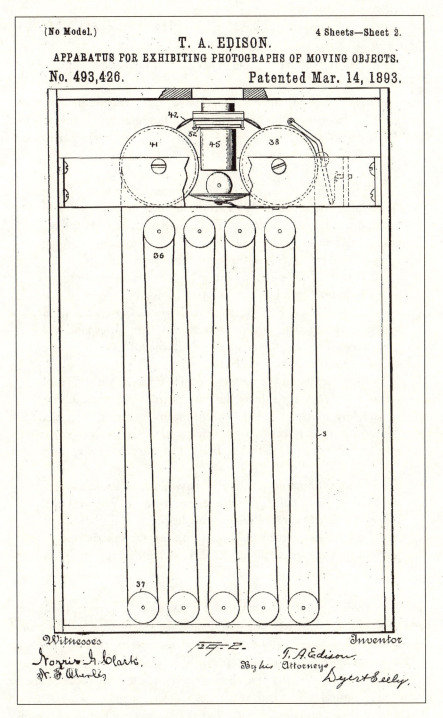

Stereo claims were filed but never used for Thomas Edison's "peep box" (1893).

notation: "Make up a stereoscopic illustration of fight to be separate from photograph. try this."[23]

The Kinetoscope patent no. 493,426, as finally issued on March 14, 1893, and titled "Apparatus for Exhibiting Photographs of Moving Objects," had several stereo claims. Claims 9 through 12 from the patent read as follows:

> 9. The combination of a film or surface having on it pictures of a moving object taken stereoscopically side by side, means for moving said film or surface rapidly forward at a regulated speed, a projecting lens or prism, and means for superimposing said pictures on said lens or prism, substantially as described.
>
> 10. The combination of a film or surface having on it pictures of a moving object taken stereoscopically side by side, means for moving said film or surface forward at a regulated speed, means for superposing said pictures, and a screen colored to correspond with the subject of the photograph onto which the superposed pictures are thrown, substantially as described.
>
> 11. The combination of a film or surface having on it pictures taken stereoscopically side by side, means for moving said film or surface forward at a regulated speed, means for superposing said pictures, and a screen colored to correspond with the subject of the photograph, onto which the superposed pictures are thrown, substantially as described.
>
> 12. The combination of the film having pictures taken stereoscopically on it in pairs side by side, means for moving said film, the light and reflector for illuminating the pictures, the heat absorbent between said light and film, and means for superposing the pictures and rendering them visible as a single picture, substantially as described.[24]

The Kinetoscope patent also included a description of color reproduction and proposed that "the reproduction of stereoscopic photographs of moving objects gives a very vivid impression of movement, and the coloring just described adds to the realistic effect."[25]

The schematic drawing labeled "Figure 4" in the Kinetoscope patent depicted an optical system with two lenses for stereoscopic viewing of moving objects. Both Gosser and Hendricks have enumerated Edison's previous patent applications for the Kinetoscope, which had numerous stereo inten-

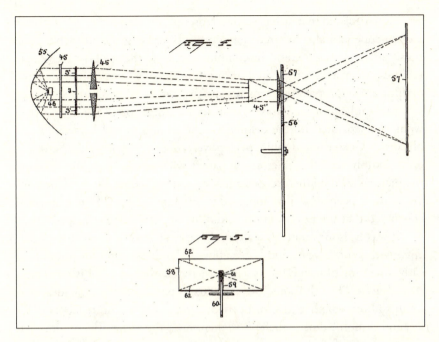

"Figure 4" of Edison's Kinetoscope patent depicted a stereo optical system (1893).

tions struck down by the examiner, who cited prior patents in France (Groualt patent no. 169,144 of May 28, 1885) and Britain (Gaetano Bonelli patent no. 1,588 of June 12, 1865)[26] and stated that "it is old to take pictures stereoscopically."[27]

Nevertheless, despite these patent claims, there is no evidence that Dickson or Edison ever successfully built a stereoscopic Kinetoscope. Nor did they ever successfully project stereoscopic motion pictures.

A curious U.S. patent (no. 588,916) issued to Willard Steward and Ellis Frost dated August 24, 1897, titled "Kinetoscope" and "Improvements in Kinetoscopes" within the body of the specifications describes a "box or casing in which a wheel or roller" is mounted around a series of pictures that are arranged with a "plurality of reflectors" displaying succeeding pictures. Figure 4 of these patent drawings displays side-by-side images and lenses that "may also be placed between the reflector and the pictures or the eye, whereby the apparent size of the picture may be changed, and by using two visual openings and properly arranged pictures or objects the usual stereoscopic effects can be produced."[28]

Writing in an 1895 issue of *Scientific American,* E. W. Scripture, a Yale professor, declared that he had seen stereoscopic moving pictures in Edison's Kinetoscope and also inferred stereo projection using the anaglyph. "The possibilities of this method in combination with the strobo stereoscopic pictures, such as are used in Edison's kinetoscope, are limitless," wrote Scripture; "by proper arrangements the two sets of pictures of the kineto-scope might be projected in two colors to the same spot on the screen. They would be separated by the colored spectacles and would be seen as real objects."[29] Gosser noted, quite perceptively, that "most probably, Scripture either simply read the kinetoscope patent with its stereo claims, mistook the pictures in the kinetoscope as having a stereoscopic effect, or perhaps saw a stereo-peep-show that appeared to be a kinetoscope-like machine and concluded that the Edison machine did, in fact, have stereo images."[30]

It was Edison's strategy to always be very aggressive in the use of pat-ents, with both caveats and applications. Hendricks cites an instance of early discernment into Edison's patent methods from an 1881 issue of *Electrical Review:* "His [Edison's] plan appears to be to patent all the ideas that occur to him, whether tried or untried, and to trust to future labors to se-lect and combine those which prove themselves the fittest. The result is that the great bulk of his patents are valueless in point of practicability; but they serve to fence the ground in from other inventors."[31]

To a great extent, Hendricks's rigorously researched book *The Edison Motion Picture Myth* was written to rehabilitate William K.-L. Dickson's reputation. "It is to the young Scotsman who came from Virginia in 1883 to seek his fortune in the great city that the American film owes more than to any other man," wrote Hendricks.[32] It was also Dickson who negotiated on behalf of Edison for cellulose nitrate film to be purchased from George Eastman and, with an order placed in 1891, determined the 35-mm format with vertical sprocket holes, which has remained the standard for motion picture film ever since.

Georges Demeny's Stereo Phonoscope

Georges Demeny was Etienne-Jules Marey's assistant at the laboratory that Marey built in the west of Paris near the Bois de Boulogne and called the "Physiological Station." Demeny was interested in filming scenes of magic, dance, and theater with stereo-chronophotography. Several surviving films provide evidence that he used stereoscopic lenses in the years 1893–1895, and depict a man sawing wood and a man sneezing, likely in imitation of Edison's "Kinetoscopic Record of a Sneeze" from January 1894.[33]

After getting a patent (no. 15,709) in Great Britain on September 1, 1892, for a device he called a "phonoscope" that combined a phenakisto-scopic disk with a synchronized cylinder phonograph, Demeny developed a stereoscopic version, which he patented (no. 12,794) on June 30, 1893. "I can also arrange two concentric circles of images on a disc," Demeny wrote in the patent, "provided with two concentric series of windows pass-ing before the eyes in such manner that the images fixed against the open-ings a and a', which are taken in stereoscopic series, are seen at the same time. I call this arrangement the zoetropic stereoscope."[34]

E. J. Wall, in an article on "Stereoscopic Cinematography," states, per-haps mistakenly, that Demeny had constructed a stereoscopic "kinetoscope" or "mutoscope."[35] Gosser notes that Wall may have been referring to De-meny's disk apparatus.[36]

Using a film advance device he called a "beater," Demeny obtained a French patent (no. 257,257), which included claims for a stereoscopic double system with twin lenses in both a camera and projector. This patent, however, may not have actually been reduced to practice.

Dickson Leaves Edison

Frustrated by Edison's refusal to develop motion picture projection for the screen, Dickson left Edison's employ in April 1895 and went into partner-ship with Elias Koopman, Harry Marvin, and Herman Casler to form the American Mutoscope Company, which eventually became known as Bio-graph. Initially, the company was known as the K.M.C.D. Syndicate, using the initials for the last names of each of the partners. Even though the Mutoscope was evolved by Dickson, patent rights for the device were writ-ten under Casler's name because, as Hendricks notes, "if Dickson had been the patentee of the devices upon which the American Mutoscope Company based its fortunes, he, because of his former association with Edison, may have been quite prevented from using these patents. The sturdy young Bio-graph organization would have been nipped in embryo."[37]

Casler's patent for the "Mutoscope," no. 549,309, granted November 5, 1895, makes no provision for stereoscopic moving images. The Mutoscope was not presented to the public until 1897, shortly after general projection of motion pictures in America had begun and K.M.C.D. had developed both their Mutograph camera and the large-format (2¼-inch) film Biograph projector that was considerably different and superior to the Edison "Vitascope" projector, which the inventor had developed with the assistance of Thomas Armat and C. F. Jenkins after Dickson's departure.

Both the Kinetoscope and the Mutoscope represented the apotheosis of the peep show tradition. Even though both devices would seem ideal for stereoscopic viewing of moving images, they were manufactured strictly for viewing of conventional "flat" pictures.

Though it is questionable whether the actual device was ever marketed, a July 10, 1900, U.S. patent, no. 653,520, by Frank Moniot and Louis Garcin of New York, proposed a "Kinetoscope Attachment for Stereoscopes," whose object was "to provide a stereoscope so arranged that it may be used for viewing pictures in the usual manner and also for viewing 'animated' pictures." Attached by hinges to the lens board of the stereoscope was a frame that carried a sliding shutter: "This shutter is designed by moving over the inner sides of the lenses to alternately open and close the view through said lenses." The frame carrying the shutter was swung upward for normal stereoscopic viewing of still images. Viewing of the animated images, alternating between the left and right eye, required "two pictures" that would "illustrate a figure in different positions."[38] The "Kinetoscope Attachment for Stereoscopes" was a variation of the Motoscope viewer manufactured by D. Appleton and Co. in 1860 and discussed in chapter 2.

Stephen Herbert sent me a stereoview card produced by the American Mutoscope and Biograph Company dated 1908. It is no. 4 in a set and titled "Give Them to Me," a standard size 3½ by 7–inch stereoview card typical of those used for arcade viewing as a sequence of still stereoscopic images. The conventional Mutoscope cards used for motion picture viewing consisted of a single image, 4 by 6 inches in size and printed on double-weight photographic paper.

Over the years, Dickson had retained his interest in stereoscopy, and in Great Britain on March 29, 1899, he filed patent no. 6,794, for a "Stereo Optical System" that was applicable to both still and motion pictures. The patent (no. 731,405) for this device was granted in the United States on June 16, 1903, with the title "Stereoscopic Apparatus."

Dickson's stereo system used two prisms, which reflected the images at right angles onto the negative. One of the prisms was mounted on a movable rod, which allowed for a variable interocular distance. It may have

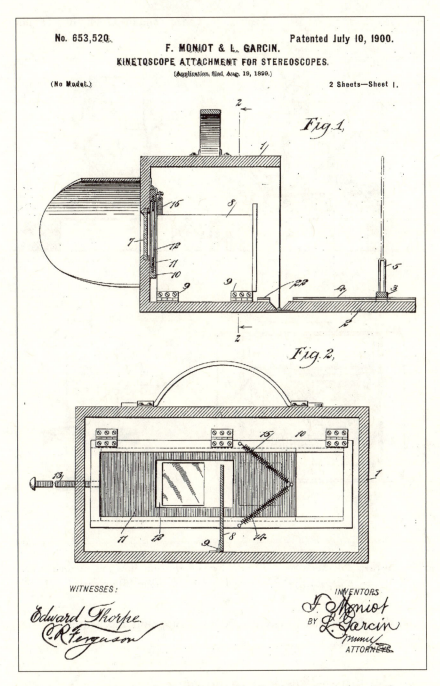

No. 653,520.

F. MONIOT & L. GARCIN.

KINETOSCOPE ATTACHMENT FOR STEREOSCOPES.

(Application filed Aug. 19, 1899.)

Patented July 10, 1900.

(No Model.)

2 Sheets—Sheet 1.

Fig.1.

Fig.2.

WITNESSES:

Edward Thorpe.
C. R. Ferguson.

INVENTORS
F. Moniot
BY L. Garcin

ATTORNEYS

"Kinetoscope Attachment for Stereoscopes," by Frank Moniot and Louis Garcin of New York, a good idea that never made it to market (1900).

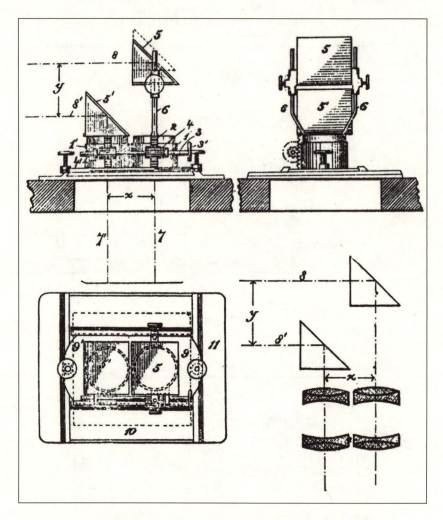

William K.-L. Dickson's Stereo Optical System, showing how the interocular between lenses (y; *upper left*) can be varied by moving prism (5) on shaft (6) (1899).

been this device that Dickson used to photograph the Mutoscope stereoview card sets.

C. F. Jenkins—Prescient Inventor

Charles Francis Jenkins was a persistent inventor, who first demonstrated his "Phantascope" film projector at the Cotton States Exhibition in Atlanta,

Georgia, in 1895. On April 2, 1901, Jenkins was granted U.S. patent no. 671,111 for a "Stereoscopic Mutoscope." This device used stereoscopic pictures that were taken in "rapid succession" and in which "the pictures of the two series—the right-eye series and the left-eye series—are made to alternate in the Mutoscope, so that the pictures of the two series are presented alternately in rapid succession." In the instant "any single picture is visible," the "view from the other eye is cut off," and the result "is that all the pictures of the two series are unconsciously combined, and the observer perceives moving figures in substantially the same relief as if the actual objects, instead of pictures, were in the field of vision."[39]

Jenkins also used alternating left-eye and right-eye images for a prior U.S. patent, no. 606,993, dated July 5, 1898, which was a "Device for Obtaining Stereoscopic Effects in Exhibiting Pictures" on an ordinary strip of film. The patent drawing depicts the optical printing of two "film-strips bearing only right pictures and left pictures, respectively, and each having its pictures approximately a picture's width apart" on to a single piece of horizontally traveling film with the left and right-eye film frames alternating.[40] For viewing of the alternate frame projection, Jenkins presciently proposed a pair of opera glasses with the left and right eyes alternating electronically via a solenoid that was also connected to the projection device.

Whether Jenkins's specific patent was reduced to practice or not, alternate-frame stereoscopic technology was to find repeated use throughout the early years of cinema and on to the present day. The 1980s saw a widespread use of alternate-frame "above/below" 35-mm single-strip stereoscopic film technology, and alternate field (video) projection is still in use in the twenty-first century.

In his 1898 book, *Animated Pictures,* Jenkins made numerous references to stereoscopic motion pictures. In that work, he clarified the 606,993 patent by stating that he intended that the two separate 35-mm left- and right-eye images were to be optically printed alternately on 70-mm or "double width" positive film. A complex binocular eyepiece was also proposed with two "reflecting prisms" for each eye, and Jenkins acknowledged the difficulties of the system.[41]

Jenkins went on to found the Society of Motion Picture Engineers (SMPE) in 1919 and developed numerous applications for mechanical scanning television in the 1920s. In the October 1920 issue of the *SMPE Journal,* then called *Transactions,*[42] Jenkins wrote about his work adapting aerial stereo-cinematography for use with geographic analysis and map-making. Though

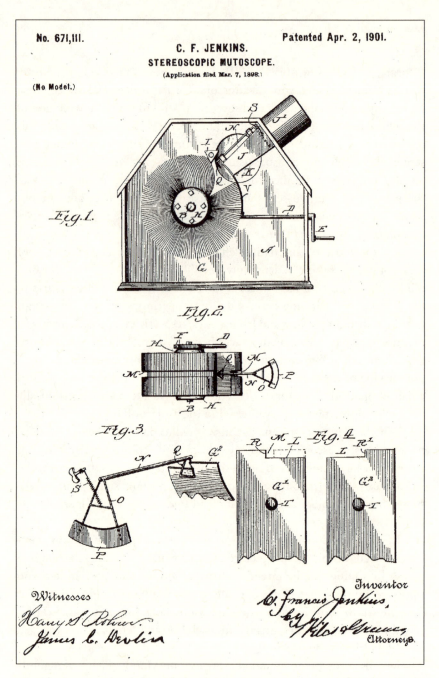

Fig.1.

Fig.2.

Fig.3.

Fig.4.

Witnesses

Inventor

Attorneys.

Charles Francis Jenkins's "Stereoscopic Mutoscope"; it is a mystery why Biograph didn't produce this, an ideal format for stereoscopic movies (1901).

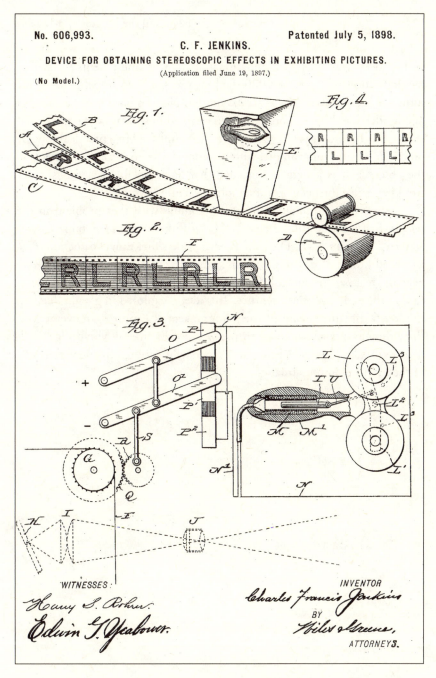

WITNESSES:

Harry S. Rohrer.

Edwin T. Yeabower.

INVENTOR

Charles Francis Jenkins

BY

Stiles & Greene,

ATTORNEYS.

Charles Francis Jenkins's "Device for Obtaining Stereoscopic Effects in Exhibiting Pictures," a prescient proposal for an alternate-frame stereo system using opera glasses (1898).

he had been granted two foundational alternate frame patents for stereo-
·scopic motion pictures, by 1920, Jenkins was less than hopeful about the
future of 3-D movies. "Stereoscopic motion pictures have been the subject
of considerable thought and have been attained in several ways," the inventor wrote, "but never yet have they been accomplished in a practical way. By
practical, I mean, for example, without some device to wear over the eyes of
the observer. It is generally conceded that its acceptable accomplishment
would enhance the beauty of the motion picture."[43]

Stereoscopic filmmakers are still grappling with that problem in the
twenty-first century. Billy Bitzer, D. W. Griffith's master cameraman, characterized William K.-L. Dickson as "the world's first movie cameraman."[44]
By an odd form of historical paradox, the growth of the movies may have
been in great part responsible for the demise in popularity of the stereoview
card in the early twentieth century. This idea is explored in greater detail in
chapter 5. But the dream of early filmmakers of capturing movement in
three dimensions is evident with the work of Eadweard Muybridge, William Kennedy-Laurie Dickson, and others who labored in the very cradle
of the motion picture's infancy.

4

Stereoscopic Projection

The Marvelous Ducos du Hauron

Louis Ducos du Hauron has been called both an "advanced prophet"[1] of the cinema and a "savant."[2] A French patent (no. 61,976) issued to du Hauron on March 1, 1864, and an addition to it granted December 3, 1864, demonstrate the inventor's understanding of practical problems involved in motion picture projection.

The original patent was very complex in using over 270 lenses in conjunction with collodion glass plates. The addition to the patent, which Gosser characterizes as "brilliant,"[3] provided for the use of a flexible band moving between two spools and carrying a series of drawings or photographs. The flexible band was moved by a row of teeth protruding from its bottom every two frames and keeping it in register for moving lenses and a viewing screen.

Walter Stainton, after much effort, acquired a copy of du Hauron's patent addition of December 3, in which the inventor wrote, "here is what constitutes the perfected apparatus which I have constructed and which works very well."[4] This language suggests that du Hauron may actually have built his moving picture machine, though many film historians are doubtful and, as Gosser suggests, "unfortunately, the materials and knowledge did not then exist to make his proposals attainable."[5]

Despite this ingenious patent and uses for it that include screen projection, stereoscopy, reverse movement, and animation, it is for his discoveries with color and the anaglyph that du Hauron will be remembered. Using complementary-colored red and blue lenses for left- and right-eye image selection is a primeval but effective way to present a 3-D image, though some people may be troubled by the color "bombardment" different to each eye that is necessary for the effect. In the mid-nineteenth century, the anaglyph proved itself as a viable means of stereographic projection. As early as 1613, the phrase "stereoscopic projection" had been used in a treatise on optics by Francois d'Aguilon, a Jesuit monk from Brussels.[6] But, according to Gosser, d'Aguilon had "used the term to mean the perception of relief rather than the reproduction of images on a screen."[7]

Quigley notes that Claude Francois Milliet de Chales, a rector at Chambery, France, had made improvements to the magic lantern of Athanasius Kircher and had "studied the eye and knew that the image is upside down on the retina. He investigated other vision problems, including angular vision and vision at long range, considered binocular vision and the images formed by each eye."[8] Without providing direct quotation, Quigley asserts that de Chales, in his monumental work of 1674, *Cursus seu Mundus Mathematicus* (Course on the Mathematical World), wrote about "the nature of color and the laws of light. De Chales even attempted three dimension projection! Even now," Quigley wrote (in 1947), "many efforts are being made to achieve 'three dimension' motion pictures without the use of special glasses or other viewing devices for the spectators."[9]

E. J. Wall suggested that, as early as 1717, attempts at stereoscopic projection evolved from experiments with colored glass and complementary colors.[10] By 1841, a German scientist, Heinrich Wilhelm Dove, had developed a subtractive printing process.[11] Complementary-colored stereoscopic methods were either additive or subtractive. With colors combined in additive, light-transmissive methods, the result is white. With colors combined in subtractive methods, for reflection, black is produced. Dove was alleged to have presented one image of a stereo pair in blue on a white ground and the other image in red. When the superimposed images were viewed with red and blue glasses, a black object on a white ground was visible in three dimensions.

Wilhelm Rollman, a German, was familiar with Dove's work and in 1853 discussed a complementary-color process for subtractive filtration in which stereoscopic drawings rendered with blue and yellow lines were viewed through red and blue lenses.[12] Rollman could have been the first to project images with complementary colors.

The first to project stereoscopic images using complementary colors in France was Charles d'Almeida. In 1858, d'Almeida placed red and green filters over two magic lanterns and provided the audience with red and green glasses. "I set out to produce an arrangement," wrote d'Almeida, where "the three dimensional images could be seen from various points in the room."[13] In the same paper, d'Almeida considered the possibility of projected stereoscopic images in motion:

> In ending this note, I do think I am occupied at this moment in discovering a simple combination which will allow me to give the movement of images and to reproduce in relief the effects of the phenakistiscope. This will be a new method of demonstration that the discovery of Wheatstone will bring to the sciences, especially in mechanics and astronomy.[14]

Alfred Molteni, an optician and manufacturer of magic lanterns from Paris, became renowned in 1890 for his anaglyphic magic lantern shows. These shows were projected with a "biunnial" magic lantern that had two vertically stacked projection lenses. Molteni attributed the invention of the anaglyphic projection process to d'Almeida.[15]

It is to Louis Ducos du Hauron of Algiers that we owe the term "anaglyph." Originating from the Greek, it translates literally as "again—sculpture." Many of the 3-D pioneers, like du Hauron, the Lumiere brothers, and Frederick E. Ives, were exploring the nature of color in photography, and it was through this work that their invention of anaglyphic processes came about. The thrust of their efforts was to create a more perfect replication of nature through photographic reproduction, and the attempts to capture both color and 3-D information in an image can be seen as an expression of that goal. Their early efforts eventually led to anaglyphic projection of motion pictures.

With a letter of 1862, du Hauron had clarified the specific nature of subtractive filtration:

> *Physical Solution of the Problem of Reproducing Colors by Photography.*—The method which I propose is based on the principle that the simple colors are reduced to three—red, yellow and blue—the combinations of which in different proportions give us the infinite variety of shades we see in nature. One may now say that analysis of the solar spectrum by means of a glass which passes only one color has proved that red exists in all parts of the spectrum, and the like for yellow and blue, and that one is forced to admit that the

Alfred Molteni's biunnial magic lantern with two vertically stacked projection lenses, used to project anaglyph images (1890) (photo by Ray Zone; Erkki Huhtamo Collection).

solar spectrum is formed of three superposed spectra having their maxima of intensity at different points.[16]

In 1862, du Hauron also described a design for a camera that made three separate negatives with a single exposure using angled transparent mirrors, a precursor of modern one-shot cameras. By exposing black-and-white negative film through filters of these colors, a three-color separation was produced. This is the underlying principle behind all color printing. In 1870, du Hauron printed a three-color stone lithograph design that demonstrated his principle.

With subtractive filtration in the anaglyph, it can be seen that the red lens sees the blue image and cancels out the red. The blue sees the red and cancels blue. Because a given color filter cancels out the same color in the photo or print it is called "subtractive."

In 1891 in France, du Hauron registered a patent (no. 216,465) for anaglyphs, which he subsequently registered in the United States on August 20, 1895, as patent no. 544,666 titled "Stereoscopic Print." Du Hauron acknowledges that "Mr. D'Almeida, a French Physician, and after him, Mr. Molteni, have obtained in a very elegant way the production of double images arranged for binocular vision and united in a single picture when properly viewed." After describing the process whereby separate magic lanterns project transparent stereoscopic images through red and green filters which the spectators view with corresponding colors so that "the perfect picture is thrown into relief as distinctly as in the best stereoscope," du Hauron clarifies that "the subject of the present invention realizes this phenomenon not merely at night and by the artificial light of a lantern, in the form of an immaterial image projected on a screen, but in full daylight and in the form of a print or photograph."[17]

The anaglyph image, as du Hauron describes it, appears as "a palpable object, which can be taken in the hand or laid upon a desk, for inspection, or hung against a wall like a picture." The reality of printed anaglyphs, du Hauron suggests, is greater than those projected:

> The effect will be so much the more striking, inasmuch as there is nothing of the appearance of a show, phantasmagoria, or entertainment of any kind about it—i.e., neither darkness, screen, nor magic lantern. The only objects visible are plainly those which belong to the ordinary every-day world, a lithograph, a drawing, or a print lying on a table.[18]

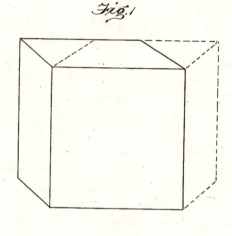

Louis Ducos du Hauron's anaglyph printing techniques were adopted quickly in the United States (1895).

Continuing his research with color photography and the anaglyph, du Hauron also worked on a method for full-color, "polychromatic" anaglyphs.

The Problematic Mr. Green

In his book on stereoscopic photography, Arthur Judge wrote:

> The first stereoscopic negative film was made by the late William Friese-Greene, who took out a patent in 1893 for his method of stereoscopic film projection. . . . This film was made with a special camera using two lenses arranged side by side; it was the first successful attempt to take a stereoscopic film of a moving picture and Friese-Greene's is the prior patent of the world for films taken and projected in this manner. It was necessary to use a viewing stereoscope in connection with the positive films thrown on a screen.[19]

A number of other sources have made similar statements. John Norling, for example, wrote:

> The first stereoscopic motion picture system was made by William Friese-Greene who patented his process in 1893. He used two negative films, one behind each lens. The positive images were projected side by side on a screen and viewed through a cumbersome stereoscope permitting each eye to see only the picture intended for it. The complexity of this system barred it from any commercial application.[20]

James L. Limbacher, citing Norling's article in his book *Four Aspects of the Film*, wrote, "William Friese-Greene, the British motion picture pioneer, patented a third-dimensional movie process before 1900. He used two films projected side-by-side on the screen. The viewer had to use an actual stereoptican [*sic*] to see the depth, which made it impractical at the time for theatrical use."[21]

These statements regarding Friese-Greene very likely derive from the book *Friese-Greene: Close-Up of an Inventor* by Ray Allister.[22] Allister's book was the basis for the 1951 Festival of Britain feature film *The Magic Box,* a biography of the inventor, featuring Robert Donat as Friese-Greene and a veritable roll call of British cinema stars, including Laurence Olivier, Michael Redgrave, and Peter Ustinov in cameo appearances. *The Magic Box* was directed by John Boulting, produced by Ronald Neame, and adapted for the screen by Eric Ambler. It was filmed in three-strip Technicolor by Jack Cardiff and shown at the Telekinema Theater especially built for the festival, which was held as a centennial celebration of British accomplishments since 1851, the year of the first exposition at which Queen Victoria examined Brewster's lenticular stereoscope and gave it her endorsement.

William Friese-Greene led a highly melodramatic life as a photographer and inventor. Continually investing his earnings in motion picture devices, color printing, and stereography, Friese-Greene experienced both wealth and numerous financial crises. When Friese-Greene collapsed and died after making an impassioned statement at a meeting of British movie producers in 1921, he was found with one shilling and ten pence in his purse, the price of a movie ticket at the time. By this time, he had filed over sixty patents in Great Britain for devices ranging from Inkless Printing, Airships, Electrical Transmission of Images, Explosives, and Games to numerous patents for color photography and motion pictures.

The British motion picture industry, atoning for their previous apathy to Friese-Greene, financed an elaborate funeral, as well as an imposing marker in Highgate Cemetery that bears the following inscription:

William Friese Greene
The Inventor of Kinematography
His genius bestowed upon humanity
The boon of commercial kinematography
Of which he was the first inventor and patentee
(June 21st, 1889 No. 10301)[23]

It was her childhood recollection of Friese-Greene's funeral that led Allister to write his biography. Though Allister remarks in her author's note to the book that the work was "the result of two years' careful research" and that "every statement can be authenticated,"[24] there are a number of questionable statements in the book regarding Friese-Greene's work with stereoscopic motion pictures.

Allister writes that Friese-Greene was experimenting with celluloid to eliminate static charges in 1888 and with Mortimer Evans designed a stereoscopic motion picture camera:

Friese-Greene wanted "real life" on the screen, and real is not flat—in a physical sense, anyway. He particularly wanted to make three-dimensional pictures. He believed he could get the stereoscopic effect by taking photographs simultaneously through two lenses and superimposing the two images on the screen during the process. The new camera was built with two lenses which could be used separately or simultaneously.[25]

Allister claims this camera was built by Chipperfield of Clerkenwell and was ready for delivery in January 1889, along with celluloid films that were thin, clear, even strips in rolls of 50 feet.

In response to this declaration, photographic historian Brian Coe writes: "The only two-lens camera with which Friese-Greene worked was that patented by Varley in 1890; Evans does not appear to have had anything to do with it."[26] The most thorough inquiry into the claims of priority by Friese-Greene for stereoscopic cinema has been done by Coe, with an article titled "William Friese-Greene and the Origins of Cinematography" published in three installments in *Screen: Journal of the Society for Education in Film and Television* in the March–April 1969, May–June 1969, and July–October 1969 issues.

Allister relates the story of Friese-Greene using the stereoscopic camera to film his cousin Alfred Carter arriving on foot in Hyde Park with his three-year-old son Bert. A sympathetic portrait of the inventor is drawn as he waits: "Friese-Greene stood beside his camera—and gazed straight ahead, seeing nothing. For the first time in his life he was sickly nervous. Here beside him was everything he had been working for during nine hard years, the mechanism he had invented, the material he had found. Supposing they failed!"[27] A portion of this stereoscopic footage survives today and is in the collection of the Science Museum of London.

This footage was actually shot with a stereo camera built by Frederick Henry Varley and patented as no. 4,704 in Great Britain on March 26, 1890. Gosser states that this camera "was the first stereo cine roll camera ever built."[28] Friese-Greene had met Varley in late 1889 or early 1890 and had acted as cinematographer using Varley's stereo camera in Hyde Park. Three years later, on November 29, 1893, Friese-Greene was granted British patent no. 22,954 for a camera that was practically identical to Varley's stereo camera.

Replica of the Varley–Friese-Greene stereo camera: exterior *(above)* and interior *(below)* (1890) (Gordon Trewinnard Collection).

Allister also tells a colorful story of the night that Friese-Greene developed the Hyde Park stereoscopic film and modified the camera to act as a projector. In his excitement, the inventor rushed out into the street and recruited a policeman as witness to the projection. Did the inventor actually achieve projection? "The surviving few frames of film exposed in such a camera, together with the design of the camera itself, show that it operated at only two or three pictures a second," surmises Coe. "There is no evidence that a satisfactory projection was ever given."[29] The film in the Varley stereo camera was 6 3/4-inch unperforated celluloid holding 3-inch square stereo images side by side.

Varley and Friese-Greene appeared before different photographic groups to exhibit the stereo camera. An account of one such demonstration is given in the *Photographic Journal* from November 21, 1890:

> Mr. F. Varley then explained a stereoscopic camera he had designed for taking successive negatives at a rapid rate. The mechanism of the instrument was shown and explained. . . . Mr. Friese Greene stated that he had made exposures with the instrument at the rate of four or five per second, and showed a long roll of exposures that had been exposed and developed. [30]

Gosser pointed out that "There was no attempt at projecting a positive at this meeting and neither of the two men claimed such a projection had ever taken place."[31]

Allister uses other colorful descriptions of Friese-Greene's attempts to create stereoscopic motion pictures. One such device was a projector intended to produce "artificial scenery" for backgrounds in theaters using stereoscopic movies in color:

> The stereoscopic effect was obtained by a double lantern with double converging lenses that threw pictures of a continuous band of film. There was a revolving cylinder between the lenses of each lantern, each cylinder being so formed as to cause the intensity of the corresponding view on the screen alternately to increase to its full extent and as gradually to diminish to nothing. When one view was at its full brightness, the other was cut off, and vice versa. But the view on the screen had always the same brightness.[32]

Varley had made some experiments with alternate-frame stereoscopic projection, which he characterized as "dazzling," and wrote about them for the *British Journal of Photography.*[33]

The slow speed of the camera/projector, however, remained an even bigger obstacle to this form of stereoscopic projection. According to Allister, Friese-Greene made another attempt at alternate-frame stereoscopic technology in 1911: "He would try another method, too, of getting stereoscopic pictures. The film was to be arranged longitudinally, pictures taken through two lenses, but alternate negatives were to be produced by light passing through only one or other of the lenses. . . . He patented both these ideas in 1912."[34]

Where did Allister get her information? Much of the anecdotal information might have come from members of the Friese-Greene family who Allister interviewed. In addition, according to Coe, "Most of these claims seem to be derived from Friese-Greene's own affidavit made in America in 1910. It was made for use as part of a legal action between the Motion Picture Patents Company and the Independent Motion Picture Company, but was never called in evidence. It is full of the most surprising inaccuracies."[35]

So where does Friese-Greene fit into the history of stereoscopic cinema? "It is certain," answers Coe, "that his work had no appreciable effect upon the introduction of cinematography."[36] Nevertheless, Coe closed his *Screen* article on Friese-Greene on a compassionate, yet discerning, note:

> The picture of the man that emerges from the papers and demonstrations at the Photographic Society is of one lacking in method, dabbling in all kinds of fruitless experiments without plan or control. . . . He was unable to express himself clearly—at least on paper; it is frequently difficult to understand from his writings what he is trying to say. . . . His ideas and suggestions were always far ahead of his ability to carry them out.[37]

Friese-Greene, who changed his name from just plain Willie Green at the outset of his photographic career, was a colorful and tragic figure in the history of stereoscopic cinema. His enthusiasm for 3-D film was boundless. But he may not have actually reduced his patents to practice.

John Anderton—Polarizing Light

The anaglyph was not the only method used to separate left- and right-eye images for stereoscopic projection in the nineteenth century. Polarized light projection was used with magic lanterns for stereo projection of slides by John Anderton in 1891.

Polarization of light is possible because light vibrates on a plane and can be made to transmit light at different angles when projected through quartz, tourmaline, crystals of calcite, and other translucent materials. Erasmus Bartholinus, a Danish physician, discovered polarization in 1669 when he was looking through a piece of Iceland spar crystal and noticed that light passing through the different facets of the crystal was split into two parts. Bartholinus designated this splitting of the light as a "double refraction," or "birefringence."[38] Later, in 1828, a Scottish physicist named William Nicols diagonally split an Iceland spar crystal and glued the two parts back together. Light entering the split crystal was transmitted in two different directions. This split spar crystal became known as "Nicol's Prism."

A British physicist named John Anderton used a modified Nicol's Prism for a stereoscopic projection system patented in the United States on July 9, 1895 (no. 542,321) as a "Method by Which Pictures Projected upon Screens by Magic Lanterns Are Seen in Relief" after having first taken out patents in Great Britain (no. 11,520) and France (no. 224,813). Anderton's projection produced a stereoscopic effect "by employing two polarizers arranged at right angles to each other and in separate lanterns, in combination with a pair of analyzers [glasses] having polarizers therein arranged at right angles to each other, the analyzer for one eye being adapted to obliterate the picture produced by one lantern while permitting the picture produced by the other lantern to be seen."[39]

The polarizers in Anderton's system consisted of a "bundle of thin glass plates" arranged so as to produce "polarization by transmission" and affording "the means to the observer of obliterating the image at will." In this manner, "a separate picture is conveyed to each eye, and as these pictures form a stereoscopic pair the impression is irresistibly conveyed to the mind of one picture only, and that a solid one." A distinct advantage of polarizing projection, as opposed to anaglyphic projection of stereoscopic images, is the fact that "colored pictures may be utilized as readily as those in black and white" without any detriment to the color values in the image. The analyzers in Anderton's system consisted of "small opera glasses or any other suitable form," and "the slides used in the lantern are prepared in the ordinary way from a stereoscopic negative—that is, a negative consisting of two pictures of the same object taken from different points of view." A silver screen is necessary for polarizing projection to maintain the specific linear axis of light for transmission or blocking of the left- and right-eye images. One of the drawings in Anderton's patent depicts the "non-depolarizing screen" upon which the stereoscopic images are projected: "Said screen may be of calico or other material having a coating of silver or other metallic elements or compound, ground glass, metal, or other suitable substance, the essential characteristic being that it is of a non-depolarizing nature."[40]

Both anaglyphic and polarizing projection of stereoscopic motion pictures would be used in the twentieth century. While polarizing projection had the advantage of depicting full-color images in stereo, the use of anaglyph projection eliminated the necessity for a silver, or non-depolarizing, screen.

From Bioscope to Alabastra

Max and Emil Skladanowsky were magic lantern showmen who inherited the business from their father, who had been creating "dissolving views" for many years. From 1892 to 1895, Max, the technical genius of the pair, created what was called the "Bioscope" projector for showing a series of photographs captured on Eastman Kodak celluloid film 44½ millimeters wide (split 89-mm stock).[41]

The Bioscope was a double projector with a speed of eight frames a second running two strips of celluloid carrying twenty pictures, each advanced with intermittent movement using a worm gear mechanism patented (no. 88,599) on the same day the brothers began public projection with their device. On November 1, 1895, the Skladanowsky brothers set up their Bioscope projector in the Berlin Wintergarten and showed a program of two children doing a dance, gymnasts working out, a juggler, and a kangaroo boxing match.[42] As early as 1903, Max Skladanowsky produced a series of anaglyph booklets with scenic views of Europe and Germany. But did the Skladanowskys ever produce a stereo Bioscope camera or projector?

In a June 29, 2005, email, historian Alexander Klein cited an article written by Erich Skladanowsky, Max's son, in the November 1955 issue of *Image and Sound* magazine. In this article, Erich stated that his father "had made arrangements for an upgrade [to the Bioscope camera] to shoot stereo." Drawings and photos of the Bioscope camera, Klein noted, "clearly show the space for a second lens and film spool; the photos show the same: the space for the second lens is closed by a piece of wood." Klein concluded, "it's most unlikely that Skladanowsky ever shot stereo movies. But based on his 1903 anaglyph experiments, he certainly had intended to shoot in stereo."[43]

An inventor named Oskar Messter had seen the Bioscope projections at the Wintergarten and decided to build his own projectors using, initially, a Geneva movement with five slots and subsequently a four-slotted Geneva cross. "It was Messter, and not Skladanowski [*sic*]," writes Gosser, "who was the most significant figure in building a German film industry."[44] Adds Albert Narath, "A process to achieve a three-dimensional screen image interested Messter from the beginning."[45] On July 23, 1896, Messter had filed a patent application for a "stereo rapid viewer,"[46] but the patent was not granted.

In 1909, Messter became intrigued with "Pepper's Ghost" images as practiced by Wilhelm Engelmann. This style of cinematographic projection dates back to 1862. The audience was actually looking through a plate of glass carrying a reflection at a 45-degree angle. The reflection was seen on a different plane and appeared to float over the background projection, and "due to dimensional differences on the film, viewers 'saw' the persons move both back and forth, so that an illusion of depth was sensed."[47] Messter created a projection system based on Pepper's Ghost which he called "Alabastra" and presented color-tinted movies with music and song accompaniment, receiving excellent notices from the press. Subsequently, at the Court Theater in Darmstadt, Czar Nicholas II of Russia and Prince Heinrich of Prussia attended such presentations.

To optimize the background projection, a Professor Thorner developed a process where a special projector incorporated a half-silvered mirror over a concave mirror to reflect the movement of the actors above the background. A patent (no. 598,712) for this process was applied for on May 19, 1932, and Messter carried out practical experiments using the system. "It could not gain acceptance," writes Narath, "because the manufacture of similar mirrors was expensive and dimensionally limited."[48]

Pepper's Ghost

In 1798, a Belgian optician named Etienne Gaspard under the stage name "Robertson" thrilled the citizens of Paris with magic lantern projections on smoke with his "Phantasmagoria" illusions, presented in an abandoned chapel of the Capuchin convent near Place Vendome. The undulating smoke reflecting projected images created a three-dimensional effect of movement in an actual environment.[49] As Erik Barnouw observes, "The magician was indeed a film pioneer. A virtuoso technician, he had already contributed importantly to the prehistory of cinema. For more than a century he had offered illusions based on projected images, which often made unsuspected use of concealed 'magic lanterns.' With these he could already do astounding things, and the Cinematographe was merely the next logical step."[50]

Sir David Brewster, the inventor of the lenticular stereoscope, speculated in his *Letters on Natural Magic* that projected images on smoke may have been used in the ancient world by priests to summon "spirits" for worshipers and invoke awe on ritual occasions. Brewster even suggested that these priests might have used mirrors to focus light in the manner of a magic lantern.[51]

The invention of limelight in 1820 provided a powerful new tool for the magician to create optical effects and illusions. In 1847, the magician Henri Robin was using limelight for his "Living Phantasmagoria" illusions, and he established his own theater in Paris—the Theater Robin—to display them. In London, the Royal Polytechnic Institution opened in 1838 and in the 1860s began to feature spectacular sideshows touted as "Wonders of Optical Science" with magic lantern illusions.

Because Professor John Henry Pepper was director of the Polytechnic during the period in which the enormously popular magic lantern spectacles were presented, the visual effects came to be known as "Pepper's Ghost Illusions." Using a very large sheet of glass tilted at a 45-degree angle to the audience, apparitions would seem to appear in the midst of live actors on the three-dimensional stage. Because the new projection techniques incorporated live actors, they were characterized as "Living" Phantasmagoria.

A civil engineer named Henry Dircks developed some of the new techniques for projection, which he explained in his 1864 book *The Ghost!*[52] Pepper's Ghost illusions were used in a stage dramatization of Charles Dickens's 1843 story, *A Christmas Carol.* The Polytechnic also presented "Temptations of Saint Anthony," with the saint visited by spectral women.

When Messter's Alabastra, sometimes characterized as "Stereoplastics," was exhibited in London in 1911, stereographer Theodore Brown became interested in the Pepper's Ghost process and created refinements for it, which he subsequently patented and renamed "Kinoplastikon."[53] A review in the March 20, 1913, issue of *Bioscope* magazine by "OUR SPECIAL CORRESPONDENT JOHN CHER" described a performance:

At the Kinoplastikon Theatre I had the pleasure of inspecting the stereoscopic moving pictures of which so much has been heard and so little seen in England. The hall is underground. The sheet hangs far back, on a level with the stage. Amongst other reels, a "Pathe Gazette" was shown. . . . Never before had I witnessed such a moving picture spectacle. It was practically the illusion of life, remarkable and astonishing, almost uncanny in its realness.[54]

Brown had applied for a Kinoplastikon patent just days before the performance witnessed by Cher. The patent was titled "Improvements in Cinematograph Apparatus for Producing Stereoscopic or Plastic Effects" and made use of a dark chamber immediately behind the glass reflection.[55] The Kinoplastikon performances took place at the Scala Theater in London, a 920-seat venue off Tottenham Court Road, which Charles Urban was also using for presentations of his "Kinemacolor" motion picture program.

On April 28, 1913, a reporter for the *Times* of London wrote: "Visitors to the Scala Theatre last week were able to see the latest development of the cinematographic art—living stereoscopic pictures shown without a screen. With the production of living pictures in their natural colours it was thought that all that was possible had been achieved, but Kinoplastikon takes us a step further."[56] Stephen Herbert points out that it is likely that Kinemacolor motion pictures were projected for the Kinoplastikon presentations. Cecil Hepworth's "Vivaphone" synchronized sound process provided aural accompaniment, using a large horn and a gramophone. An advertisement for the Kinoplastikon program heralded "An Amazing Mystery Seen for the First Time in This Country, Living, Singing, Talking Cameo Pictures in Plastic Relief without a Screen." Programs handed out at the Scala Theater presentations trumpeted the Kinoplastikon as "Singing, Talking, Moving Picture Figures in Solid Stereoscopic Relief, without a Screen."[57]

Despite its lively, short-term success in 1913, Brown's Kinoplastikon did not continue to develop as a part of the motion picture landscape. Pepper's Ghost, however, remains in use in the twenty-first century at theme parks such as Disneyland in Anaheim, California, with the "Haunted Mansion" attraction and at the Gene Autry Western Museum in Los Angeles.

Stereo Projection in France

When Alfred Molteni projected his anaglyph lantern shows in Paris in 1890, much interest was generated on the part of his scientific colleagues.

In a handwritten letter to Jules-Etienne Marey dated December 5, 1890, Molteni explained the process:

> A double apparatus is used, consisting of two lanterns which have been rendered convergent so that they project on to the same point on the screen. One of the sides of a stereoscopic image on glass is placed into one of the lanterns; the second image goes in the other lantern. Behind one of the photographs there is a red glass, behind the other a green glass. Thus there is both a green image and a red image on the screen. If the viewer now puts on a pair of spectacles with the green glass over one eye and [the] red glass over the other, as each eye sees a distinct image, a sensation of three dimensions is created immediately and is very pronounced in some images.[58]

Early in 1896, an inventor from Lyon named Paul Mortier created a "reversible" stereoscopic cinematographic camera that he called the "Alethoscope" and patented it in France on February 17, 1896. "Coloured filters in complementary colours must be placed in front of the projection lenses," wrote Mortier, "and the spectators must be provided with similarly coloured spectacles (anaglyphic process)."[59] Mortier also specified a complex system with glasses using alternating shutters driven by electric current. The glasses were synchronized to the shutters running in two interlocked magic lanterns, with everything running off alternating current. A small multipolar alternator kept everything synchronized.

August Rateau, a Frenchman residing in London, developed a similar process in 1897 using two cameras recording alternate left-right views and projecting them onto a screen; a shutter in the projector was synchronized for viewing with binocular-style goggles.[60]

Claude Louis Grivolas, an electrical engineer in France, was interested in cinematic illusion. In working with French film pioneer Charles Pathe in 1897, Grivolas attempted to build a system that used forty frames per second and at the same time began to experiment with stereoscopic motion pictures. Two of Grivolas's 1897 French patents (no. 263,574 and no. 266,131) were for a combination camera and projector device that used parallel lenses with over a 4-inch separation and rotating shutters that alternately eclipsed the left- and right-eye images.

Grivolas improved his system in 1901 (French patent no. 310,864 and British patent no. 10,695) by optically printing the twin-strip stereo photography onto a single strip of film for projection with the left- and right-eye views alternately printed. His projection machine used red and blue color filters and a four-part shutter with two opaque and two transparent sections. The audience used red and blue glasses to view the alternate-frame anaglyph motion picture.[61]

Watershed Years—1895–1896

The year 1895 marked a watershed for motion pictures. It was the year that motion pictures, in the form that the twentieth century experienced them, were first publicly exhibited. First, the Skladanowsky brothers exhibited the Bioscope films on November 1 at the Wintergarten Theater in Berlin. Then, on December 28, the Lumiere brothers gave the first public exhibition of Cinematographe films at the Grand Café in Paris at 14 Boulevard des Capucines.

During the spring and summer of 1894, Edison's Kinetoscope was installed in penny arcades, hotel lobbies, and amusement parks across the United States. After licensing Thomas Armat and Francis Jenkins's Vitascope projector and renaming it the "Vitagraph," Edison's agents, Norman Raff and Frank Gammon, booked Koster and Bial's Music Hall for the public debut for a program of films on April 23, 1896.

By this time, technical standards for motion picture production and exhibition had begun to settle. Gosser has summarized the five essential technological design concepts:

1. Flexible 35-mm film with Edison standards for double perforations
2. A running speed according to Lumiere standards of sixteen frames per second
3. A prominent loop ahead of the film gate to eliminate tearing of the film during transport
4. Intermittent movement through a synchronized shutter
5. A light source[62]

Edison failed to patent his Kinetoscope in Europe, and his technology was "reverse engineered" by Robert W. Paul and Birt Acres in Great Britain. Both men proceeded to build a motion picture camera and a projector called the "Theatrograph" according to the technical standards that were being established.

By 1897, film production and exhibition had begun to proliferate as the motion picture industry was established. Cinema proper entered into a short-lived novelty period that lasted up to 1905. It was a time when the public was entranced to see any "living picture" or "animated photograph" with subject matter of any kind. By 1905, the number of Nickelodeon theaters and purpose-built motion picture halls was rapidly increasing as the movies began to tell stories, marking the end of cinema's novelty period. The stereoscopic film, however, remained in a novelty phase, and would continue to do so for another five decades.

5

Cinema's Novelty Period

"Stereopticon" Entertainment

Did the evolution of film narrative in the early years of the twentieth century and increasingly sophisticated production techniques in some manner diminish the drive to convey depth on the motion picture screen? Was the inherent "flatness" of the movie canvas, in X and Y parameters only, an advantage to filmmakers somehow in telling a story?

When projection of motion pictures for the public had been achieved in 1895, the novelty period for conventional cinema commenced and lasted for a brief period. Film historian Charles Musser has given us a concise definition of the novelty period of cinema: "Despite many stable elements the cinema underwent a staggering array of fundamental changes between 1895 and 1907. During the first months of widespread projection, short (one-shot) films were enjoyed primarily for their ability to reproduce lifelike motion and exploit isolated presentational elements."[1]

In a highly influential essay, Tom Gunning characterized the period before 1908 as a "cinema of attractions," which "by its reference to the curiosity-arousing devices of the fairground" denotes "early cinema's fascination with novelty and its foregrounding of the act of display":

> The objects of this display varied among current events (parades, funerals, sporting events); scenes of everyday life (street scenes, children playing, laborers at work); arranged scenes (slapstick gags, a highlight from a well-known play, a romantic tableau); vaudeville performances (juggling, acrobatics, dances); or even camera tricks (Melies-like magic transformations). But all such events

were absorbed by a cinematic gesture of presentation, and it was this technological means of representation that constituted the initial fascination of cinema.[2]

This fascination with the realism of the projected image lay at the heart of the magic lantern exhibitions that preceded and overlapped the beginnings of motion picture presentation. Consequently, there was a general confusion on the part of the press and public as to the two modes of representation. C. Francis Jenkins, in his 1898 *Animated Pictures* book, delineated their similarities: "The fact is, the moving picture machine is simply a modified stereopticon or lantern, i.e. a lantern equipped with a mechanical slide changer. All stereopticons will, sooner or later, as are several machines now, be arranged to project stationary pictures or pictures giving the appearance of objects in motion."[3]

Musser notes that the Langenheim brothers of Philadelphia began selling photographic lantern slides in wooden frames of 3⅝ by 6⅞ inches in the 1840s. The introduction of photographic images to magic lantern projection at this time created great excitement on the part of the press. "By magnifying these new slides through the magic lantern," wrote one editor, "the representation is nature itself again, omitting all defects and incorrectness in the drawing which can never be avoided in painting a picture on the small scale required for the old slides."[4]

The use of the term "stereopticon" to describe two-dimensional projections, Musser suggests, may have been a result of the fact that a single image from a stereo pair was frequently printed as a magic lantern slide: "Since the Langenheims and other dealers sold stereoscopic views and lantern slides, they often cut the double images in half and projected individual slides with a magic lantern. Because these slides were so frequently used in the lantern, Americans often called the projector of photographic slides a 'stereopticon.'"[5]

When John Fallon, a chemist from Massachusetts, introduced improved stereopticon projection in 1863, P. T. Barnum quickly appropriated the device for his American Museum in New York and presented an exhibition of photographic views under the title of the "Great English Stereopticon." Musser notes that "the journalistic praise accorded the stereopticon evokes the amazement that greeted the first screen images and anticipates the later enthusiasm for the novelty of projected motion pictures."[6]

Publicists and the press, with hyperbole and promotion for planar en-

tertainments in the nineteenth century, misrepresented them as stereoscopic presentations. One writer, extolling the virtues of "Cromwell's Stereopticon," wrote that "it presents us with a literal transcription of the
actual. . . . Stereoscopic pictures are placed before us which are the exquisite
shadows of the photograph, freighted with all the minute details of the
subject as it really exists, not a flat monochromatic shadow, but a rounded,
glowing picture, thrown up into splendid relief with all its marvelous ac-
curacy magnified."[7]

During the novelty period of cinema, magic lantern projections, useful
for musical interludes and reel changes, would continue to be interspersed
with motion pictures. The stereopticon entertainments were increasingly
organized sequentially with what Musser characterizes as "spatial continu-
ities" and "the creation of a spatial world."[8] Many of the one-shot films of
the novelty period for motion pictures, though monocular, exploited a
radical use of perspective and point of view. The first movies were trying to
convey a stereoscopic "feeling" on the flat motion picture screen.

Phantom Rides

The earliest films of the novelty period of cinema exploited perspective and
the sense of depth by having moving objects rush toward or away from the
camera. Known as "actualities," these films were the first documentaries
and recorded real events at sixteen frames per second as they were happen-
ing. Parades and trains were prominent subjects because they made dramatic displays of motion in perspective.

Gerry Turvey has discussed the use of depth in the early British actuality films from 1895 to 1901. Parades and processions were typically shot on
the oblique to accentuate depth: "The filming of street parades regularly
adopted a position shooting along the length of a road, thereby gaining a
clear perspective effect and a sense of depth to the image." Depth of field
and juxtaposition of foreground and background were also important com-
positional elements. "Camera placement and the view aesthetic also entailed attending to the foreground-background aspects of picture
composition," Turvey notes. "Perspective views, diagonal constructions,
movement towards the camera and movement of the camera all drew attention to composition 'in depth' and the depth of field clarity of early cinematography could be used to advantage in actuality picture composition."[9]

The "view aesthetic" in novelty films implied the presence of the audience and its point of view. Gunning notes that "the attraction directly addresses the spectator, acknowledging the viewer's presence and seeking to quickly satisfy a curiosity. This encounter can even take on an aggressive aspect, as the attraction confronts audiences and even tries to shock them (the onrushing locomotive that seems to threaten the audience is early cinema's most enduring example)."[10]

When the Lumiere brothers began to exhibit their films at the Grand Cafe in December 1895, the novelty of motion toward the camera created a strong impression, and one of the films in particular, titled *L'Arrivee du train,* frightened the audience with its moving image of a locomotive heading straight toward them. In a famous essay titled "The Kingdom of Shadows," the Russian author Maxim Gorky, described the effect of this historic moment:

> Suddenly something clicks, everything vanishes and a train appears on the screen. It speeds straight at you—watch out! It seems as though it will plunge into the darkness in which you sit, turning you into a ripped sack full of lacerated flesh and splintered bones, and crushing into dust and into broken fragments this hall and this building, so full of women, wine, music and vice.[11]

Working as a cameraman at the Biograph Company, Billy Bitzer shot a similar film called *Empire State Express.* In his autobiography, Bitzer cited a *New York Times* review of a program of actuality films that played at the Olympia Theater in New York in 1896. "The finest of all these pictures was one of the Empire State Express going at sixty miles speed," wrote the reviewer:

> The train is seen coming out of a distant smoke cloud that marks the beginning of a curve. The smoke puffs grow denser on the vision, and soon coach after coach whirls to the front, and it seems as though the entire left hand section of the house [the Olympia] would soon be under the wheels. . . . The cheers that greeted the picture and its representation were as great as those for [President] McKinley.[12]

For his research, Turvey drew on the descriptions in the motion picture catalogues of the producers of the early actuality films. "These descriptions lay out the ideas and concepts underpinning actuality practice and are often accompanied by mention of a particular visual 'effect' being sought by the filmmaker," he noted.[13] Although movement of subjects toward the camera was often used, it was movement of the camera itself that increasingly characterized what producer's catalogues described as "stereoscopic effects." Turvey observes, "So, just as the actuality discourse associated 'perspective effects' with movement towards the camera, it associated 'stereoscopic effects' with the mobile camera—though both sets of effects were concerned with how a sense of depth and three-dimensionality might be achieved in the cinematographic image."[14]

Increasingly, however, it was movement of the camera forward through space that typified the cinema of attractions. Near the end of conventional cinema's novelty period, it was the "phantom ride" film that bridged the early actuality films with the new narrative mode of cinema. These films were generally shot from the front of a moving train with the cameraman on the front of the cowcatcher of the locomotive. "These movements too," notes Turvey, "were associated, time and again, with notions of stereoscopy."[15]

British cinema pioneer Cecil Hepworth worked in the genre. "Then around 1901, we came to a definite milestone in the shape of the *Phantom Rides* which became tremendously popular about this time," he wrote. "These were panoramic pictures taken from the front of a railway engine traveling at speed." Hepworth had to find a suitable camera, noting that "it was no use tackling that job in fifty-foot driblets." As a result, Hepworth "determined to construct a camera big enough to take a thousand feet of film at a time and take no chances. What eventually emerged was a long, narrow, black box, rather like a coffin standing on end."[16]

Robert C. Allen in an essay titled "Contra the Chaser Theory" observed that it was "easy to underestimate the power of what I call Kinesthetic films, those designed to give the illusion either of being in the path of a moving object or of actually moving through space."[17] Allen cited a film reviewer for the *New York Mail and Express* who wrote about a Phantom Ride film called *Haverstraw Tunnel* that was produced by the Biograph Company and photographed by Billy Bitzer: "The way in which the unseen energy swallows up space and flings itself into the distances is as mysterious and impressive as an allegory. . . . One holds his breath instinctively as he is swept along in the rush of the phantom cars. His attention is held almost with the vise of a fate."[18]

Bitzer's *Haverstraw Tunnel* was exhibited in a unique exhibition format called "Hale's Tours." Hepworth characterized the format as "an ingenious scheme" and noted that Hale's Tours gave the Phantom Ride film a new "lease on life":

A number of small halls all over the country were converted into the semblance of a railway carriage with a screen filling up the whole of one end and on this was projected from behind these panoramic films, so that you got the illusion of traveling along a railway line and viewing the scenery from the open front of the carriage. The illusion was ingeniously enhanced by the carriage being mounted on springs and rocked about by motor power so that you actually felt as though you were traveling along.[19]

Invented by retired Kansas City Fire Chief George C. Hale, patented on September 19, 1905, and first exhibited publicly as "Hales's 'Pleasure Railway'" at the St. Louis Exposition of 1905, the unique showcase for Phantom Ride films was very popular for a short period, with as many as 500 installations running at one time in the United States. Hale's Tours bridged the late novelty period for conventional cinema with the era of the Nickelodeon and nascent narrative cinema.

Camera movement on the z-axis in conventional cinema, even though not actually stereoscopic, has traditionally conveyed a sense of depth—an enveloping of the spectator in projected imagery moving on the screen. These two-dimensional cinematographic techniques were so effective in conveying a feeling of depth that, to some extent, the need to create a genuine stereoscopic cinema may have been diminished.

Acknowledging the Viewer

As the narrative mode began to develop in conventional cinema, around 1903, the practice of actors looking at the camera and acknowledging the spectator came to an end. As Gunning has noted, this had been one of the primary traits of the cinema of attractions and implicit in a mode of display. However, the use of point of view, as well as imagery moving toward the spectator, is also inherent in a stereographic image.

Edwin S. Porter's 1903 film *The Great Train Robbery,* produced for the Edison Company, was a landmark in establishing narrative conventions. With straight cuts between fourteen scenes, dispensing with the prior practice of using dissolves or titles between scene changes, and with fast-paced action, it became a massively popular film.

Poised on the divide between the novelty era and that of the new narrative mode, however, it included a cinematically atavistic final shot of a

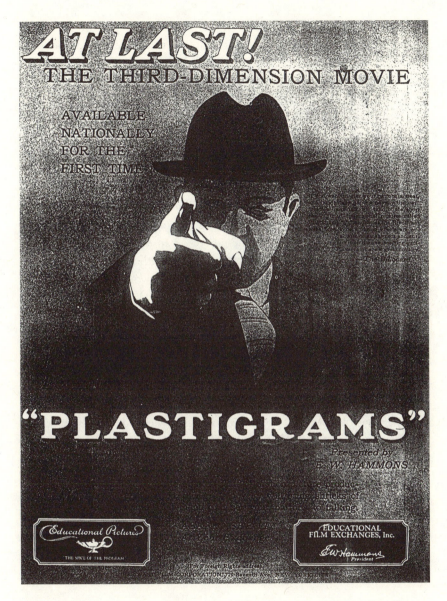

Acknowledging the viewer; advertisement from *Motion Picture News* (1924).

close-up of one of the outlaws raising a pistol and firing it directly at the viewer. This scene in an *Edison Films* Supplement was described under the heading "REALISM" as "a life size picture of Barnes, leader of the outlaw band, taking aim and firing point blank at each individual in the audience. (This effect is gained by foreshortening in making the picture.) The resulting excitement is great." The potential exhibitor was also advised that "this section of the scene can be used either to begin the subject [the film] or to end it, as the operator may choose," although it most often was used as a final thrilling coda at the end.[20]

In his essay "An Unseen Energy Swallows Space: The Space in Early Film and Its Relation to American Avant-Garde Film," Gunning wrote: "A direct look at the camera was later seen as sabotaging the developing space of narrative cinema, and it became taboo. As an offscreen glance became one of the ways shots were linked together and synthetic space was created, a glance like this directed at the audience/camera would undermine these connections."[21] Gunning cited a 1910 *New York Dramatic Mirror* article written by a film critic named Frank Woods, writing under the nom de plume "The Spectator," who commented that "picture acting is far more convincing and effective when the players appear to be ignorant that there is a camera taking their picture. Facial remarks directed at the camera destroy the illusion of reality."[22]

As motion pictures emerged out of the novelty period, the spectator became invisible, peering unseen and unacknowledged into the world of the narrative (diegesis) told onscreen. Whereas conventional cinema began to adopt this narrative technique around 1903, it would be many years before stereoscopic motion pictures would begin to do so. The motif of an actor firing a gun at the viewer would recur in a variety of stereographic media, including lenticular parallax stereograms, lithographic stereoview cards, and anaglyph motion pictures of the 1920s. Stereoscopic cinema would remain a cinema of attractions—foregrounding display and acknowledging the spectator—for decades to come.

Caligari in Utopia

Film historian Noel Burch has characterized classical Hollywood cinema with its narrative conventions as the "institutional mode of representation," or the IMR. In his highly theoretical book, *Life to Those Shadows,* Burch stated that the IMR obviated any need for stereoscopic motion pictures.

"Let me anticipate the results of my investigations and reflections," wrote Burch. He continued: "I believe it was this aspiration to three-dimensionality that was satisfied by the blossoming of the Institutional Mode of Representation around 1910, and that the latter continues to satisfy it more than all the ephemeral re-appearances of red-and-green or polarizing spectacles, raster screens, etc."[23]

The aspiration to three-dimensionality Burch traced back to nineteenth-century efforts "towards the realization of a perfect illusion of the perceptual world" and the "Recreation of Reality" to be found in both trompe l'oeil painting and photography, as with the work of Louis Daguerre, who excelled at both. Burch cited Moigno who, after the success of the stereoscope at the Great Exhibition of 1851 in London, wrote: "Photography, reinvigorated, perfected and crowned by the stereoscope, is so superior to its former self that the day will soon be here when all photographic pictures . . . will come . . . in pairs to reproduce in all their truth, in all its gentle or harsh beauty, immaterial and living nature."[24]

Burch characterized the attempts of the early inventors of motion pictures as "supplying a deficiency" in synthesizing motion, color, and relief for the "reconstitution of reality." Like alchemists of the middle ages, the inventors, with their "Stereofantascopes, Phasmatropes and Omniscopes" were seeking to find "the Great Secret of the Representation of Life." Burch called his utopian ideal the "Frankensteinian Tradition."[25] And, for many journalists of the time, the capture of motion in living photographs represented a form of triumph over death.

In a chapter titled "Building a Haptic Space," referring to the psychology of the sense of touch, Burch recapitulated the strategies by which the IMR substituted symbolic, visual depth for real stereoscopic images on the motion picture screen. First, cinema had to adopt Renaissance perspective, "the institutional representation of space," which was "in a sense a recapitulation of the decades of work which went into the constitution of monocular perspective in painting."[26]

Burch noted five characteristics of early cinema that accounted for the visual flatness of films made before 1906:

1. A more or less vertical illumination suffusing the whole field in front of the lens with a completely even light
2. The fixity of the camera
3. Its horizontal and frontal placement

4. The very widespread use of painted backdrops
5. The placing of the actors, always a long way from the camera, often spread out in a tableau vivant, all facing front, and without axial [z-axis] movement of any kind[27]

The staged films of Georges Melies, in particular, were characterized by this proscenium-based visual style. "The whole visual history of the cinema before the First World War," observed Burch, "thus turns on the opposition between the 'Meliesian' affirmation of the surface and affirmation of depth already implicit in [Lumiere's] 'Arrivee d'un train a La Ciotat.'"[28]

The 1914 Italian epic *Cabiria* was one of the first films to make use of camera-tracking shots. Director Giovanni Pastrone stated in an interview that "the camera movements were used to create stereoscopic effects. Focusing proved very difficult. All the more so in that I emphasized the impression of relief by sometimes using curved rather than straight tracks."[29]

Burch also cited Hugo Munsterberg, a pioneering popular psychologist, whose 1916 book, *The Photoplay: A Psychological Study,* contained a chapter on depth and movement in the cinema. In 1915, Griffith's landmark film *The Birth of a Nation* had appeared and dramatically illustrated the narrative possibilities for motion pictures. "To begin at the beginning," wrote Munsterberg, "the photoplay consists of a series of flat pictures in contrast to the plastic objects of the real world which surrounds us." But Munsterberg stopped at once to observe that "we have no right whatsoever to say that the scenes which we see on the screen appear to us as flat pictures."[30]

After a brief discussion of the principles of the stereoscope and stereoscopy, Munsterberg questioned "whether the moving pictures of the photoplay, in spite of our knowledge concerning the flatness of the screen, do not give us after all the impression of actual depth." After giving consideration to anaglyphic projection as well as monocular cues for the perception of depth such as apparent size, perspective relations, and shadows, Munsterberg struck at the heart of the question. Despite the use of monocular cues for depth in motion pictures, Munsterberg observed that "we are never deceived. We are fully conscious of the depth, and yet we do not take it for real depth," as a result of the fact that "we are constantly reminded of the flatness of the picture because the two eyes receive identical impressions." The experience is a "conflict of perception" in which visual depth is created through "our mental mechanism."[31]

Munsterberg's observations would seem to be borne out by a review of

The Cabinet of Dr. Caligari written by Willard Huntington Wright, "recognized both in Europe and America as one of the foremost authorities on painting and aesthetics." The review appeared in the September 1921 issue of *Photoplay Magazine* and was titled "The Romance of the Third Dimension."[32]

With quite florid language, Wright expounded on the startling effect of *Caligari* as deriving from "the first sight of land in a motion picture new world—the eastern shore of the continent which has been the quest of every Columbus of the brush—farthest east of that Arcadia of vision, the Land of the Third Dimension." Characterizing the "chief problem of the art of oil painting" as the "achievement of the third dimension," Wright declared that "mere perspective has never been enough." As a champion of modern art, of which *Caligari* was a chief cinematic example, Wright claimed that modernists had "recently discovered how to produce the effect of a third dimension" by studying and experimenting with "the laws of optics, the mutability of related masses, the fluctuability of lines, the functioning elements of tones and colors," and other factors.[33]

Hyperbolic but misguided, mistaking vision-in-the-round on a flat surface for binocular stereopsis, Wright was attempting to suggest that modernist painting movements like Impressionism and Cubism incorporated stereography. In a footnote, Burch stated that Wright's "confusion is not incomprehensible" and, comparing the film to Expressionist art and its concomitant use of primitive art, observed that "'Caligari's' imagery plays constantly on a carefully sustained ambiguity. The film's notorious visual style presents each tableau as a flat, stylized rendering of a deep space, achieved by a design of dramatically oblique strokes."[34]

By 1920, the visual grammar of classical Hollywood cinema, the IMR, had begun to be very well established. Audiences had learned to denote depth in moving images even though they did not actually perceive them using binocular stereopsis. Or, as Munsterberg had written, "We certainly see the depth, and yet we cannot accept it."[35]

Burch, with considerably greater complexity, reaffirmed what Munsterberg had observed about the mental nature of depth perception. He stated that "the continuing audience preference for the imaginary three-dimensionality of institutional screen-space over the illusionistic three-dimensionality of stereoscopic films, incline one to believe that the term 'illusion of reality' is a malapropism masking the evidence of 'a rationally selective system of symbolic exchange.'"[36]

Display versus Story

When the first publicly exhibited stereoscopic motion pictures were shown in 1915 at the Astor Theater in New York, Lynde Denig, a reviewer for *Moving Picture World,* wrote, "These pictures would appeal first by reason of their novelty, then because of the wonderful effects obtained, and after that, when they had become familiar, there would be the same old demand for an interesting story."[37]

Contemporary film theory has addressed the issue that stereoscopic display may undermine the narrative in motion pictures. With a Bachelor of Arts thesis delivered May 1, 2002, at the University of Chicago titled "The Reality of Vision with Depth: 3-D Cinematic Spectatorship," Lauren Kroiz wrote, "In the terminology of poststructuralist film theory, 3-D film foregrounds the scopophilic pleasures of cinema, while opposing traditional narrative film's model of spectatorship which is predicated on viewers' narcissistic identification with characters in the film's diegesis."[38]

There is also, through the use of a viewing device, an awareness of the body, what Laura Schiavo called "physiological optics," in the experience of viewing a stereographic image. "Clearly the stereoscope was dependent on a physical engagement with the apparatus that became increasingly unacceptable," wrote Jonathon Crary in *Techniques of the Observer,* "and the composite, synthetic nature of the stereoscopic image could never be fully effaced." The physical space a spectator occupies in relation to a stereographic image retains a subliminal presence throughout the visual experience. For spectacle and the "pure perception" of modernism, Crary noted that "the eventual triumph of both depends on the denial of the body, its pulsings and phantasms, as the ground of vision."[39]

The spectator of a stereoscopic motion picture, as a result, may not achieve the disembodied sense of invisibility that is common with a merely referent two-dimensional display. "By wearing the glasses the viewer, rather than entering into the world of depth that 3-D seems to offer, is instead bodily positioned as an observer barred from identification within the spectacle," wrote Kroiz; "positioned in filmic space as a spectator, the viewer is unable to enter film as a character, unable to identify with the characters in the narrative."[40]

Despite misleading advertising, exploitation of monocular depth cues, and the development of narrative conventions, stereoscopic motion pictures would continue to be developed by inventors and producers in pursuit of a utopian dream of realistic representation. This dream as realized on the motion picture screen, however, would continue to foreground spectacle and display, finding expression primarily as a cinematic novelty.

6

Public Exhibition of 3-D Films Begins

Developments around 1900

As public motion picture exhibition began to take place in Europe and the United States, numerous inventors proposed a variety of devices for capture and display of moving images in three dimensions. An inventor named Hymmen, in a kind of technological throwback to an earlier era, was granted a patent (no. 24,804) in Great Britain in 1897 for a spinning drum to be used in conjunction with a stereoscope for viewing three-dimensional moving images.[1]

Gosser has enumerated many of these devices, which he characterized as "interesting but generally less influential."[2] Most of these inventions were very likely only developed to the stage of a prototype. A British instrument maker named Frederick Wenham, for example, created a series of ten stereoscopic images for viewing on two phenakistoscope disks. With a letter of 1895, Wenham claimed to have built the invention in 1852.[3] As with the stereoscopic moving images of Sellers and Muybridge, Wenham arranged a sequence of still stereo images to simulate motion. Similarly, in 1865, Henry Mayhew, a British photojournalist, had persuaded his brother to create a series of six stereoscopic images. The stereo pairs were mounted on a rotating paddlewheel device for stereoscopic viewing as moving images.

Stereoscopic disk projection systems were developed as late as 1897 with the work of Thomas Cunningham Porter in Great Britain. Porter's device, patented in Britain (no. 12,921), used an electromagnetic shutter in the stereoscopic glasses that were linked to the projector and alternately projected left- and right-eye images. In 1898, the Lumiere brothers, after introducing stereoscopic still technology to the photographic market, in

France patented (no. 305,092) a polygonal plate glass device with double rows of stereo images making intermittent movement and viewed in 3-D through a shutter.

Stereo cameras to capture motion on film, however, became prevalent after 1900. Frank Donisthorpe, son of the inventor Wordsworth Donisthorpe, took out a British patent (no. 1,483) on January 21, 1903, for stereooptic devices for use with twin cameras and projectors. Donisthorpe proposed anaglyphic projection for his twin-strip system.

Carl Schmidt and Charles Dupuis were issued a French patent (no. 331,406) on April 21, 1903, for a twin-strip film system using an electromagnetic shutter with which left- and right-eye images alternated through the use of a revolving mirror. Spectators were provided lorgnette stereo glasses affixed to the chair that were connected by wires to the projector. This system, as well as Porter's, foreshadowed Laurens Hammond's Teleview process that would debut in 1922 in New York.

Shortly after the turn of the century, inventors were also working on stereoscopic attachments for existing cameras that could produce three-dimensional pictures. Wilhelm Salow, of Elberfeld, Germany, for example, was issued an American patent (no. 840,378) on January 1, 1907, for a "Stereoscopic Attachment for Photographic Cameras." Salow's attachment incorporated "two rectangular equilateral prisms" and was used for still cameras to "produce two pictures consecutively in the camera, with the distance between the pictures equal to the average distance between the eyes." The advantage of Salow's device was that it allowed stereoscopic photographs to "be taken in a perfect manner by means of an ordinary photographic camera."[4]

Aloys Wayditch of New York was granted a U.S. patent (no. 1,071,837) on September 2, 1913, for an "Attachment for Taking Cameras for Making Stereoscopic Moving-Picture Films." Wayditch's device consisted of "stationary side-mirrors placed at an angle to the vertical longitudinal center-plane of the camera" with "parallel mirrors placed back to back intermediately between the side-mirrors." An adjustment provided for moving the "side-mirrors into proper angular position relatively to the vertical center-plane of the camera." The attachment alternated the left- and right-eye images on a single strip of film. Such a mirror device prefigured many of the dual-band 3-D camera rigs of the early 1950s, which used mirrors and a micrometer for adjustment of converging optical axes.[5]

Wayditch was issued a subsequent U.S. patent (no. 1,276,838) on Au-

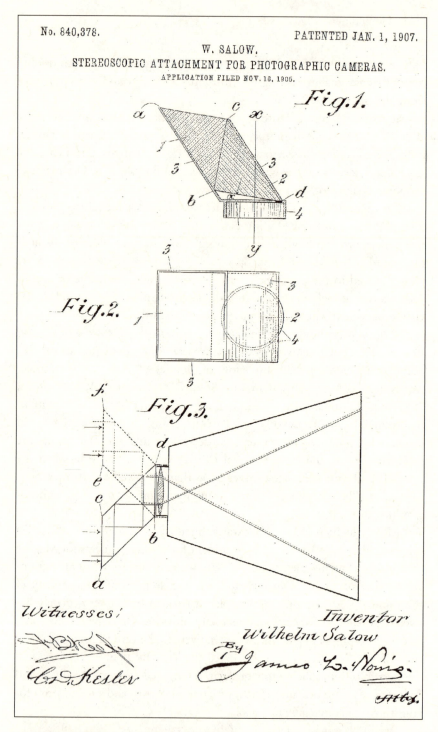

Beam splitters like those from Wilhelm Salow, "Stereoscopic Attachment for Photographic Cameras," eventually became a mature 3-D technology (1907).

gust 27, 1918, for a "Kinetoscope for Projecting Stereoscopic Moving Pictures," which referenced his 1913 patent and described a device "composed of two convex lenses in a common tube" in which, "for the purpose of projecting stereoscopic pictures, the front lens is removed [and] only the rear lens is retained."[6] Four convex lenses, instead, were used as the front lens, which was used with a shutter to project two pairs alternately.

In 1910, a compact, single-strip, stereoscopic motion picture camera was developed in France (patent no. 414,159) by Jules Richard and Louis Joseph Colardeau. The patent for this "Stereoscopic Cinematograph" was granted in the United States (patent no. 1,209,498) on December 19, 1916. On a single band of 35-mm film, two lenses, spaced 65 mm apart, were combined with two tetrahedral prisms, "which give two properly adjusted images on the band itself."[7] The left- and right-eye images were side by side on the film. Only two of these cameras were ever built, and one of them is housed today in the Musee Francais de la Photographie. Made of polished walnut, brass, and steel, the compact stereoscopic cinematograph measures 14 by 43 by 23 centimeters.[8]

The use of alternating left- and right-eye frames with a shutter in the viewing device became a commonplace idea after the turn of the century. Michael Francis Sullivan's July 4, 1916, U.S. patent (no. 1,189,308) for a "Stereoscopic Projection Apparatus" was described as an "improved apparatus for alternately cutting off the vision from the right and left eyes of an observer so that suitably projected motion pictures, or two stationary stereoscopic views projected alternately, will be seen in stereoscopic projection by the observer." The shutter mechanism for Sullivan's apparatus was "in the form of a pair of electromagnets having U-shaped cores," which gave the advantage of "thus insuring that each circuit will be closed slightly in advance of the opening of the other" and preventing the shutter from being "in a central or neutral position" where "both apertures would be partly open."[9]

Similarly, John Henley of Charlemont, Ireland, secured a U.S. patent (no. 1,284,673) on November 12, 1918, for a "Stereoscopic Projecting Machine" with left- and right-eye images taken separately and "alternately arranged upon a length of film in a stereoscopic manner" and "viewed through means arranged to alternately permit one eye to have a clear vision while dimming the view of the other eye." Henley describes motion picture stereo cinematography with two cameras "run synchronously" and "fitted on top with crosswire finder so that the axes of both may be directed accurately to the same particular part of the principle subject." The twin-strip stereo negatives were then to be "printed on the same positive film in such a manner that the pictures alternate with each other." Each spectator was provided with glasses on a "light metal rod" with an "occulting apparatus" running from a "light metal electrical vibrator."[10]

As motion picture exhibition evolved into the age of cinematic narrative, the utopian dream of three-dimensional moving images with sound was still alive. Film historian Earl Sponable, writing in the April 1947 issue of the *Society of Motion Picture Engineers Journal*, wrote about A. Manuelli, who in 1908 obtained a French patent (no. 386,737), having "as its object a bicinematographic photophonic machine for public and private displays adapted to insure fixedness of projection, stereoscopic effect, photographic reproduction of sound, etc."[11] Manuelli described a complicated machine that used three separate films.

Lucien Bull—High Speed Stereoscopy

Working as assistant director at the Marey Institute in Paris, circa 1903, Lucien Bull developed a unique system for high-speed stereo cinematography capturing 2,000 frames per second. The system was used for scientific purposes and captured the detailed movement of insects in flight. Bull created a "spark drum camera" that replicated the continuous motion of 35-mm film. Using an electromagnetic shutter, two side-by-side films were exposed and wound around drums inside a camera built from wooden frames. A metal base held the octagonal-shaped camera body steady during photography.

The first single-lensed iteration of Bull's device would only capture the insect's flight as a silhouette. "In order to obviate this drawback," wrote Frederick A. Talbot, "Monsieur Bull introduced a stereoscopic system, wherein two lenses are mounted side by side before the film box, with two spark-gaps in the same circuit. This enables two sparks to be produced simultaneously with each interruption of the primary circuit, to give two images of the object upon the sensitized celluloid films."[12]

It was necessary to build a unique stereo shutter system. According to an explanation by Talbot:

In order that the two exposures should be made upon the films exactly at the same time, a special type of shutter had to be evolved, whereby the exposure apertures were opened simultaneously at the critical moment when the cylinder commenced to revolve, because, as the films were traveling continuously, there was no necessity for an alternate and closing shutter movement.[13]

To make his system work, Bull also had to invent a clamp to hold an insect such as a fly, doing it no harm, until it was released for stereo photography. "This clamp was introduced in the stereoscopic shutter," notes Talbot, "and as the clamp releases the fly the stereoscopic shutter is opened."[14]

With a 1973 article in the *Journal of the Society of Motion Picture and Television Engineers,* Pierre Mertz wrote that Bull continued his work with stereoscopy for many years.[15]

Stereoscopic Movies in Natural Color

In her book on William Friese-Greene, Ray Allister noted the inventor's perennial attempts to create color stereoscopic cinematography: "By the end of the year 1892 Friese-Greene was working at two ideas in bliss and torment. He still wanted three-dimensional images on the screen: and he wanted pictures in the color of Nature."[16]

It was an endeavor that the inventor would pursue during his entire life. An October 1909 issue of *Scientific American* featured an extensive article with photographs about Friese-Greene's "stereo-chromo-cinematograph" with twin lenses that "are alternately exposed by the shutter." The complicated device used "a disk of glass divided into three equal sized sectors, each representing one of the three fundamental colors [red, green, and blue], and revolving it in front of the lens as the exposures were being made, each section of the sensitized film . . . being exposed through one of the three color filters." A complex "method of working color films of two lenses alternately" was used. The films ran at double speed, thirty-two frames per second. For projection, "the colored disk was again revolved in such a manner that the same relationship of picture to color filter that prevailed during the exposure of the negative was repeated."[17]

The unnamed "English correspondent" responsible for the article noted that "at the time of writing some twenty films had been secured" by Friese-Greene, "depicting varying scenes in natural life from incidents to busy streets to studies of insects and growing flowers. Not only are the color effects obtained," observed the author, "but they are stereoscopically projected." There is no further description of the means by which the films were to be viewed with binocular stereopsis, using either glasses or a mirror device. The author then added: "The effect produced upon the screen is precisely the same as that obtained with ordinary stereoscopic photographs when observed through the hand instrument devised by Oliver Wendell Holmes."[18] One can only wonder what, if any, device was used by the writer for stereoscopic viewing of Friese-Greene's natural color films.

To further confuse the matter, the author then described a "highly important development" of Friese-Greene's in which "the color filters are disposed on an endless band of transparent celluloid in the order of red, green, and blue," which are synchronous with and running in "dead juxta-position" to the sensitized panchromatic film. The author claimed that "the continual cutting in and out of the colors will enable them to be blended so easily and rapidly that the brain sees the heliochromic image only."[19]

Both color techniques described made use of the additive process. On October 19, 1909, Friese-Greene was granted a U.S. patent (no. 937,367) on the basis of his invention for color cinematography, the "Kinemato-graphic Apparatus," which he described as using "a translucent screen-band or color filter divided into successive series of primary colors" that was "disposed between an orthochromatically sensitized film and the camera lens." Provisions for twin-lens stereo photography were described in the patent: "When it is desired to take stereoscopic negatives, or to reproduce stereoscopic effects, a color screen is used with each lens, the screens being preferably so arranged that the primary colors of the two screens do not correspond at any one exposure." Friese-Greene also proposed that "the screen of one lens may be colored with alternate divisions of red and green, while the screen of the other lens is divided alternately into divisions of yellow and blue."[20] No provision for stereoscopic viewing of the color films was described.

Walter Speer, managing director of the Montpelier Theater in Brighton, England, in partnership with Friese-Greene had formed a company, Biocolour Limited, in 1911 to show films using the inventor's process.

In 1906, Charles Urban, working with G. Albert Smith in Britain, had invented the Kinemacolor process, which used red and green filters in the revolving wheel. Bernard Jones, a scientific author, described projection of Urban's "Natural Color" cinematography:

In the projecting machine, at the moment when the red filter is opposite the lens, a monotone image taken through the green filter will be in the gate and be projected, and vice versa. The images following in this order at the high speed of thirty-two images per second, the combined effect upon the screen will be a picture reflecting not only red and green, but also their complementary or accidental colours intermixed with many other hues resultant from the blending of the red and green proper.[21]

After patenting the process in Great Britain in 1906 and launching the commercial career of Kinemacolor in Great Britain in February 1909, Urban introduced Kinemacolor in the United States in December of that same year. The process proved very successful on an international level, and in 1912, Urban debuted his two-and-a-half-hour masterwork on the Delhi Durbar to King George and Queen Mary at the Scala Theater in London.

In 1912, "Biocolour Limited was forced to close down because Charles Urban and George A. Smith obtained an injunction against it," wrote Allister. "Urban applied for an injunction to restrain Biocolour from making two-colour films on the ground that the two-colour process, using green and red, was an infringement of Smith's 1906 patent." On August 22, 1912, the application for the injunction was granted. However, with financial backing from S. F. Edge, a famous racing motorist, Friese-Greene was able to challenge the injunction. When the case was appealed to the House of Lords, the Kinemacolor patent was held to be invalid because Smith failed to specify exact colors but, instead, "he conveyed that no colours of nature would be excluded and that any two colours would produce the result, which was not true."[22]

Standard film histories have neglected to mention that it was Urban who first initiated the litigation over additive natural color cinematography. Here is Terry Ramsaye's recap of the situation:

> The success of Kinemacolor inevitably attracted attack. A suit on the patent was brought by William Friese Greene [sic], the perennial British claimant to film honors. Urban won the fight through the lower courts and lost at last on an appeal to the House of Lords, on the pin-point technicality that the patentee had failed to specify the colors used in the process with sufficient accuracy.
>
> The decision was of no profit to Friese Greene. It threw the Kinemacolor process open to the world. The Kinemacolor method became in consequence the basis of practically all subsequent color processes.[23]

In his history of British cinema, *The Film Business,* Ernest Betts, after noting the "brilliant success" of Urban's Kinemacolor at the Scala Theater with the Delhi Durbar film, wrote that "it was followed by a court action which William Friese-Greene brought against him for infringement of his patent. The case went to the House of Lords and Urban lost it. Nor did Friese-Greene gain anything by it."[24] Similarly, Roy Armes noted Urban's triumphs of 1912, and then wrote, "But a law suit brought against him by his rivals (who included Friese Greene) was eventually decided against him, with the result that Urban went back to the United States during the First World War and his British company faded."[25]

Despite the demise of Kinemacolor, Friese-Greene's Biocolour process failed to thrive, in part because of the onset of World War I. Other inventors, however, persisted in working out methods for additive color stereo cinematography.

Mathias J. Vinik of New York, New York, was issued a U.S. patent (no. 1,218,342) on March 6, 1917, for a "Finder for Cameras," which was a dual-band stereoscopic color system for motion pictures. The patent drawing depicts two reels of film and two lenses contained within a single unit "embodying a reflecting shutter located between the film or sensitized medium and the lens of the camera." Vinik claimed that use of the device as described would result in "production of monochromatic films stereoscopically" and that it would provide "polychromatic pictures stereoscopically" with "suitable color filters" that were "rotatably mounted in front of the [lens] apertures." Each of the three filters—green, red, and blue—would produce a succession of pictures, and "every fourth picture of which has been taken through the same color filter or screen" would "produce a projection on the screen of a picture in natural colors provided suitable color filters are again made use of."[26] No provision for stereoscopic viewing of the two three-color films was described.

In order to design single-strip color stereoscopic cinematography, North H. Losey of Indianapolis achieved the height of complexity with U.S. patent no. 1,291,954, dated January 21, 1919. Three pages of drawings accompanying the patent were practically indecipherable, with overlapping figure numbers and lines. Titled "Cinematography Apparatus," Losey's invention "embraces means" whereby "the pictures shall appear simply stereoscopic, in other cases as approximately in natural color and stereoscopic." The photographic instrument of Losey's design is "provided with means for operating and guiding a film horizontally in order to attain

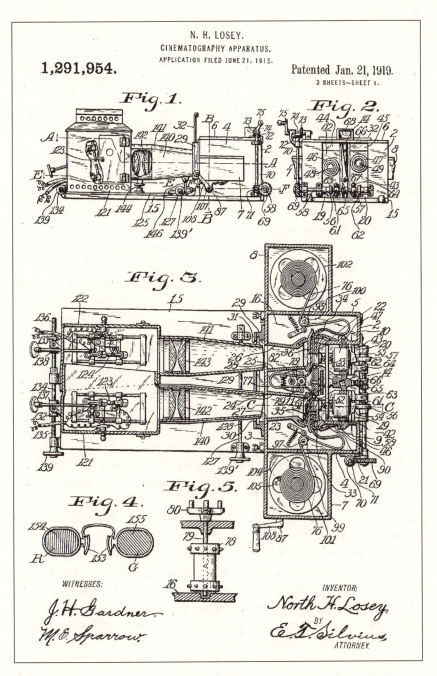

North H. Losey's inscrutable patent, "Cinematography Apparatus," reaches the height of complexity; fig. 4 illustrates the lorgnette pair of stereoscopic viewing glasses (1919).

the most satisfactory stereoscopic effect with a pair of lenses properly spaced apart, and means for operating the film so that there shall be a loop formed thereby between the two of each pair of images on the film."[27]

It was also possible to use Losey's camera as a projector, as it was "enabled to project a stereoscopic pair of images side by side, or to converge them to a common point so that the images shall be superimposed." Though Losey did not use the term, he specifically described anaglyphic projection when referring to "a pair of images" that are "projected superimposedly through a pair of color lenses and the superposed picture be unitedly viewed by means of a complementary color pair of glasses." With figure 4 of the patent drawings, Losey represents a lorgnette pair of stereoscopic viewing glasses.[28]

It is not known if the stereoscopic systems of Vinik and Losey were ever actually reduced to practice. Additive color technology, however, would certainly play a part in the first exhibition of stereoscopic motion pictures in the United States.

Edwin S. Porter—Artistic Mechanic

Edwin S. Porter (1870–1941) is primarily known as the director who filmed and edited *The Great Train Robbery* (1903). With its use of innovative editing, this film helped establish cinema as a storytelling art and announced the era of narrative motion pictures. Porter made many films and is instrumental in motion picture history. Employed by Thomas Edison, Porter manned the Vitascope projector at Koster and Bial's in New York on the night of April 23, 1896, when the very first movies were projected on a screen in the United States.

What is not generally known is that Porter, working with William E. Waddell, also made stereoscopic motion pictures. These 3-D movies might have been the first to be projected on a motion picture theater screen in the United States. Adolph Zukor, one of the founders of Paramount Pictures and a movie pioneer who inaugurated feature-length films in the early years of the motion picture, hired Porter to work at his Famous Players Film Company in 1913. In his 1953 autobiography, *The Public Is Never Wrong*, Zukor wrote about his years working with Porter: "Porter, was, I have always felt, more of an artistic mechanic than a dramatic artist. He liked to deal with machines better than with people. In a way it was his mechanical

Edwin S. Porter, 3-D filmmaker and director and editor of *The Great Train Robbery* (1903).

imagination which had caused him to improvise the story technique in *The Great Train Robbery*."[29]

On the morning of June 10, 1915, at the Astor Theater in New York, Porter and William E. Waddell screened for a private audience stereoscopic tests using the anaglyph method. The screening was reviewed in the June 26, 1915, issue of *Moving Picture World* by film critic Lynde Denig. It was also covered by an unidentified correspondent in the June 16 issue of the *New York Dramatic Mirror*. "The pictures marked a distinct advance over any of the kind made in the past, and many in the audience appeared to regard them as the forerunners of a new era in motion picture realism," wrote Denig. "Just how soon the invention will be used in the production of pictures for the regular market has not been decided, although Mr. Porter believes that the day of such a development may not be far off." Porter spoke to trade paper representatives before the screening, at which time "he made clear his confidence in stereoscopic photography."[30]

With a brief introduction before the screening, a spokesman "announced that the long sought after third dimension in photography had been gained." Porter and Waddell were stated to have conducted 3-D film experiments over a period of ten years. "Naturally, Mr. Porter does not care to divulge in detail the process by which he produces stereoscopic films suitable to standard projection machines." Three separate reels were shown. The first consisted of rural scenes, hills, valleys, houses, and figures on a country road in stereoscopic relief. "The effect is marvelously real," observed Denig. Interior scenes of "figures moving in a studio" followed along with passages from *Jim the Penman,* a forthcoming Famous Players feature film. The last 1,000-foot reel consisted of scenes of Niagara Falls.[31]

"The audience at the Astor Theater was frequently moved to applause by the beauty of the scenes which gave one the impression of looking at actual stage settings and not the shadowy figures of the ordinary picture," wrote the correspondent for the *New York Dramatic Mirror;* "trees and shrubbery stood out boldly . . . and the effect, to one accustomed to the ordinary pictures, cannot be described. The branches of trees, for instance, have the mystifying appearance of standing out from the screen and hanging over the stage."[32]

Some of Denig's criticism suggested that a dual system for projection was used with the twin-strip films, either tinted red and green or projected through colored filters and running slightly out of synchronization, with resultant phase error:

> Judging from the first samples, it may be surmised that the inventors are meeting difficulties in catching fast action. The poorest scene in the exhibition was of an elaborate Oriental dance in which the performers were blurred and the film in its entirety shimmered, something like a reflection on a lake. Then there were other instances in which quick movements failed to register.[33]

R. M. Hayes, in his book, *3-D Movies,* wrote that these films were produced as "three one-reelers in single strip anaglyphic duo-color."[34] Writing about Porter's 3-D filmmaking, Adolph Zukor recollected differently: "He used two cameras, just as two or more are used now [1953], and threw pictures on the screen by means of two projectors. He had made a lorgnette with red glass for one eye and green for the other. Seen with the naked eye, the pictures were a hopeless swirl. The lorgnette gave them three dimensions."[35] Color motion picture processes at the time largely consisted of hand-dyed film or the additive procedure with rotating color wheels, as with Kinemacolor, that was used with black-and-white panchromatic film. Several color processes at the time did use separate lenses on the camera.

In *Motion Picture Making and Exhibiting*, a book that was published a year before Porter and Waddell's stereoscopic film demonstration, John B. Rathbun wrote: "Since the ordinary two color motion pictures are often taken with a double lens camera it is sometimes possible to obtain stereoscopic effects with colored glasses as one lens only takes greens and the other, reds. This is most prominent with the use of alternate projection."[36]

Direct color photography was in its infancy, however, and it would be four years before Technicolor was to perfect its cemented positive two-color process. So it was very likely that the anaglyph one-reelers were projected through red/green filters using two interlocked projectors, as Zukor states. Porter certainly would have been capable of machining interlock devices for both stereoscopic cameras and projectors.

Frederic E. Ives—Stereo Polymath

Frederic E. Ives was a true stereo polymath, an originator of color 3-D photography, halftone printing, the lenticular parallax stereogram, and anaglyphic motion pictures. The career of this unique inventor bridged the nineteenth and twentieth centuries and brought new realism to images presented to the public.

The nature of color and its reproduction were a passion for Ives. As early as 1881, he received U.S. patents for halftone printing processes, and at the Novelties Exhibition in Philadelphia in 1885, he exhibited printing made with trichromatic halftone plates. His first color patent, no. 672,573 (July 22, 1890), described a three-color additive process to project a full-color image from positive transparencies onto a screen with a lantern. As early as 1878, Ives had been studying color photography; by 1892, he had perfected his monocular Photochromoscope. But his 3-D color triumph was the creation of the stereo Kromskop system with U.S. patent no. 531,040 (December 18, 1894).

Ives's Kromskop created the left- and right-eye stereo record on a single panchromatic glass plate using red, violet, and green filters. The Kromskop viewer was an elegant device that used transmitting reflectors and filters for stereo enjoyment of the photography. The completed Kromogram stereo views consisted of three glass plates held together by silk tapes and fan-

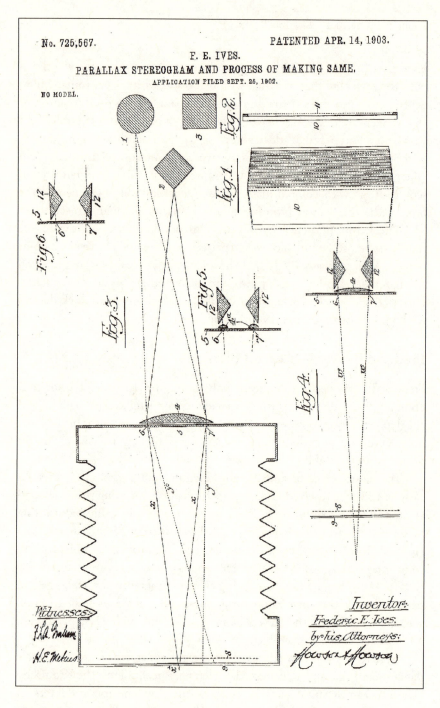

No. 725,567.   PATENTED APR. 14, 1903.

F. E. IVES.
PARALLAX STEREOGRAM AND PROCESS OF MAKING SAME.
APPLICATION FILED SEPT. 25, 1902.

NO MODEL.

The classic autostereoscopic patent of F. E. Ives, the "Parallax Stereogram and Process of Making Same" (1903).

Classic motif of a gunman firing at the onlooker; F. E. Ives's parallax stereogram, "The Brigand" (1903).

folded for storage. In 1898, Ives made a trip through Europe, producing color stereoviews with his Kromskop camera.

On April 14, 1903, Frederic Ives was granted U.S. patent no. 725,567, describing the "Parallax Stereogram and Process of Making Same." This was a lenticular autostereoscopic process, and it represents the first really practical method of producing a stereophotograph that did not require glasses for viewing. In his patent, Ives described a "photograph consisting of a composite image, in juxtaposed lines, of the elements of an ordinary double stereogram, exposed to view through a screen of alternate opaque and transparent lines, so adjusted as to give a stereoscopic effect by the parallax of binocular vision."[37] It was the first U.S. patent of its kind and would prove to have a great influence.

After the Ives patent was granted, the Scientific Shop in Chicago, Illinois, published five of his Parallax Stereograms as stock images available to demonstrate the stereo effect of the process. Typical of these images was "The Brigand," an 8 by 10–inch black-and-white glass stereogram in an oak frame and shipping box. This work depicted a cowboy pointing a revolver directly at the viewer. Within a few years, "photo-change" postcards also began to appear that created a "two-flip" animation using a plastic lenticular screen.

Ives's work with autostereoscopic imaging would be continued by his son, Dr. Herbert Ives, who in 1932 was granted a patent for a parallax panoramagram process for autostereoscopic cinematography.[38] Ives also continued working with color processes for motion picture photography and was granted many patents. His work with Jacob Leventhal proved instrumental to the first wave of stereoscopic motion pictures during the 1920s in the United States.

7

A Wave of Stereo

Anaglyph Proliferates

As the 1920s commenced, much work was being done to create color photography for motion pictures, and these technologies directly led to the use of anaglyph, which by then had been perfected for printing in newspapers and advertising supplements.[1] Color halftone printing of stereoview cards had proliferated in the early years of the twentieth century, with new manufacturers like the American Colortype Company in New York and Chicago. There was a widespread sense of a new visual world opening up, with both color and stereography, and there was highly imaginative innovation taking place.

When Louis Ducos du Hauron presented his 1893 paper on *The Art of Anaglyphs,* he stated that he would give up the rights to his anaglyph patent if anyone were "to print and publish an anaglyph image of the moon suspended in space."[2] This challenge was met and first published in the pages of the *Illustrated London News* on March 8, 1921. These were the first anaglyphs to appear in any British illustrated newspaper. The moon anaglyph, after first being made into an autochrome photographic print by Leon Gimpel, subsequently appeared in the Paris weekly *L'Illustration* on January 26, 1924. The anaglyphic image of the moon was a hyperstereo photograph made by M. C. le Morvan of the Paris Observatory. The first exposure was made on January 2, 1902, and the second on February 28, 1904. An interval of two years was necessary for proper alignment of the moon, and its formation was "as it would appear to a giant with eyes 28,125 miles apart, and observed by him from a distance of 240,000 miles."[3]

In America, the printed anaglyph had been appearing in Sunday newspapers as printed inserts since the late 1890s. Alfred J. Macy, with U.S. patent no. 1,386,720 (August 9, 1921) titled "Stereoscopic Picture," built on du Hauron's earlier patent by describing "a stereogram which shall cause an object or a portion of an object to appear to be nearer the observer than the surface of the stereogram." Macy also described an anaglyphic projection whereby "the amount of its protrusion can be predetermined" and production of "a reference plane" in the stereogram obtaining "the impression that certain parts of the picture are in front of the surface."[4]

Macy trademarked his anaglyph process as "Macyart," and in the 1920s and 1930s, through the American Colortype Company, he licensed anaglyph printing for promotional brochures to companies such as the Frigidaire Corporation and the Mohawk Rubber Company.

Shadowgraphs

Laurens Hammond of New York, in addition to originating the Hammond Organ, was also an inventor of stereoscopic displays and processes. Hammond had the ingenious idea to create stereoscopic shadows of live performances using red/green rear illumination and a backlit screen. David Hutchison writes that "in 1918, the Keith-Abbey vaudeville circuit featured a 3-D shadow-graph act, which featured a chorus line strutting its stuff and flashing high kicks at the audience. In the midst of the number, a translucent screen was lowered in front of the girls who were then illuminated from the rear with red and green light—casting red and green shadows on the screen."[5] The stereoscopic shadowgraph act was a popular addition to theatrical productions in the 1920s, and it had a significant influence on the anaglyph films that were subsequently produced.

Hammond's U.S. patent no. 1,481,006 (January 15, 1924) was titled "Process of and Apparatus for Stereoscopic Shadowgraphs." Like Macy's patent, Hammond's described stereoscopic imagery that projected into the viewer's space with negative parallax. Hammond described "stereoscopic shadows of actual objects such as actors" that the spectator sees "moving about in the hall over the heads of the audience and in advance of the stage on which they are actually located."[6]

Hammond licensed the process to Florenz Ziegfeld, who incorporated it into the "Ziegfeld Follies," his "National Institution Glorifying the Glorious American Girl," and staged it as the "Ziegfeld Shadowgraph,"

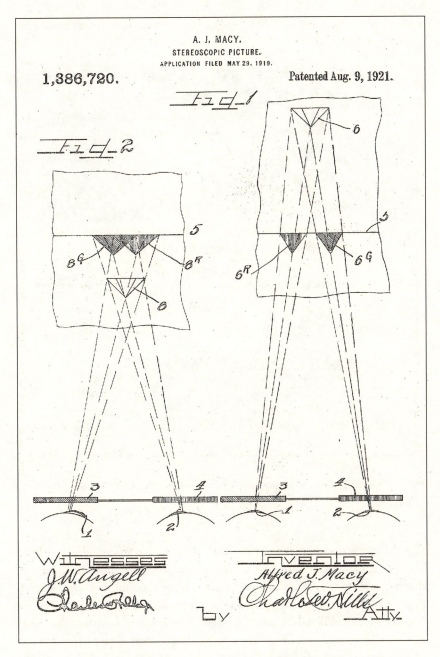

Alfred J. Macy's system for creating anaglyphs projecting out in space, titled "Stereoscopic Picture" (1921).

Florenz Ziegfeld's Shadowgraph with "Follies-Scope" glasses; playbill for the
Ohio Theater (1925) (Ray Zone collection).

viewable with red/cyan anaglyphic "Follies-Scope" lorgnette glasses. Ham-
mond was credited for the "Optical Illusion," and his patent number was
dutifully printed in the program adjacent to the description of the "Shadow
Number." The shadowgraph had a long successful run in Ziegfeld's Follies.
It played at least till 1925, when it opened on January 4 at the Ohio Theater in Cleveland. The Ohio Theater playbill included tipped-in Follies-
Scope glasses and an admonition to the audience that "SOILING THE
GLASSES IN ANY WAY WILL SPOIL THE EFFECT ENTIRELY."[7] The shadow num-
ber played between live acts. Other players featured on the program were
comedian Bert Wheeler and motion picture actresses Billie Burke and Lina
Basquette.

Teleview—The "Eclipse" Method

Hammond's Teleview was a twin-strip 3-D motion picture system that debuted at the Selwyn Theater in New York in December 1922. Teleview used a twin-strip 3-D camera with two lenses 2⅝ inches apart, dual projectors, and a revolving electrical shutter affixed to the armrest of each spectator's seat. Though Teleview received considerable press and good notices, it was used for only one theatrical run at the Selwyn Theater.

"Stereoscopic motion pictures have come," declared the film critic for the *New York Times* in the regular "Screen" column of the newspaper after an advance preview screening in October 1922. "The impression of objects in relief that one gets from them is startling, so real that it seems inappropriate to speak of it as an illusion." Devised by Hammond and electrical engineer William F. Cassidy, who were described by the *Times* as "two rather young Cornell graduates,"[8] Teleview, according to the *Motion Picture News,* was shown "through double projectors which are placed in such a way as to throw the two pictures on the screen, first one and then the other in such rapid succession as to make practically one picture."[9]

Hammond was issued a number of U.S. patents that protected the Teleview system. His first (U.S. patent no. 1,435,520) was dated November 14, 1922, and titled simply "Stereoscopic Motion Picture." In it Hammond pointed out that the normal sixteen frames per second standard for motion pictures required that the screen be darkened, with the projector shuttering at forty-two times or more per second to eliminate visible flicker on the screen. Hammond's solution was to use a rotating shutter in the viewer running at 1,500 rpm. The rotating shutter was an aluminum plate "about the thickness of a piece of paper" driven by a "self-starting synchronous motor of the three-phase alternating-current type."[10] The electric motor itself was only 1¾ inches in diameter to fit easily within the individual viewer.

His second (U.S. patent no. 1,506,524), dated August 26, 1924, and titled "Stereoscopic Motion Picture Device," provided a more detailed description of Teleview and a schematic drawing of the shutter viewer. A third (U.S. patent no. 1,658,439), dated February 7, 1928, and titled "Stereoscopic Picture Viewing Apparatus," specifically addressed the shutter viewer integrated into the retractable seats that were commonly found in movie theaters of the time. In this patent, Hammond described the shutter viewer as "a heavy, substantial, not easily broken instrument which it will not be necessary to place on the nose or hold in the hands."[11]

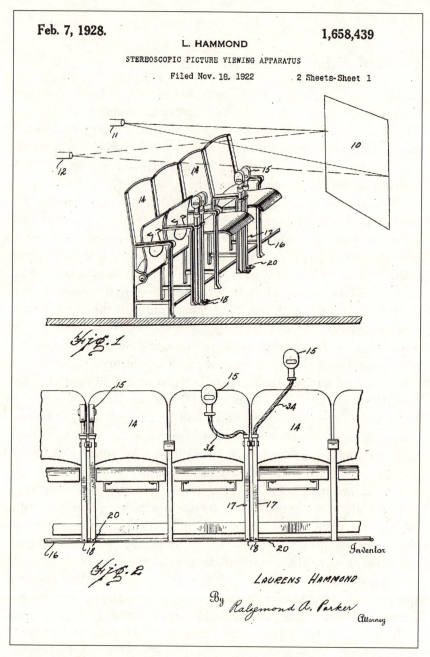

Laurens Hammond's Teleview, showing individual shutter viewers on a flexible goose neck at each seat (1922).

The notice of the Teleview program that appeared in the *New York Times* film review column, "The Screen," displayed the headline "Vivid Pictures Startle." According to the unidentified reviewer: "Those who went to the Selwyn last night were surprised, sometimes startled and often delighted with the vividness of the pictures they saw and the unusual effects obtained by the use of the Teleview device." The program began with a number of stereoscopic drawings. A jug marked "Rye" was extended out into the audience, and "the temptation was to reach out and seize it. A dragon stretched his weird form over the orchestra pit." A number of stereoscopic still pictures of the Canadian Rockies were followed by motion pictures of Hopi and Navajo Indians, "all startlingly realistic in three dimensions."[12]

Hammond contrived a way of using the Teleview process to display his shadowgraphs for the next segment of the program. "It was not a motion picture at all, but a shadowgraph dance, performed by real people behind a screen. When viewed through the teleview the shadows were not flat, as they would be ordinarily, but rounded, and separated as figures from each other. The effect," noted the reviewer, "was decidedly novel and pleasing." The Teleview program concluded with a feature film titled *M.A.R.S. or Radio-Mania*, a dramatic effort with Grant Mitchell and Margaret Irving in the leading roles. Stereo cinematography was by George Folsey, who would later shoot *Forbidden Planet*, another science fiction film, in 1955. *M.A.R.S.* was "drawn out to a tedious length," observed the reviewer, "and burdened with much dreary humor in the subtitles. It illustrated the use of the third dimension in a photoplay, however, and the part of its action laid on the planet Mars, as imagined in the hero's dream, permitted a number of bizarre effects in costumes and settings."[13]

The critic was generally positive about stereoscopic viewing of the images in Teleview: "It does not seem, in the first place, that they are trying on the eyes. . . . Also, the faces of people sometimes seem more clear in them than on the usual screen." A comment was made that "the pictures are reduced in size," but that "although the close-ups are clear, the length of any limb or object extended toward the camera in them is grotesquely exaggerated." Concluding, the reviewer wrote that it had been "a most interesting experience, with undetermined future possibilities."[14]

M.A.R.S. may have seemed overlong to the critic, but it seems to have been a result of bad writing and not the stereoscopic visuals. *M.A.R.S.* was an isolated motion picture achievement, but it was not the first stereoscopic feature film.

The Power of Love

The September 30, 1922, issue of *Film Daily* ("the Bradstreet of Filmdom") reported "Stereoscopic Pictures Shown" with a preview of the motion picture *The Power of Love,* the first 3-D feature film, at the Ambassador Hotel Theater in Los Angeles. "Scenes of the Yosemite valley were shown as a prologue," it noted, as well as the fact that "the film received continuous applause."[15]

Popular Mechanics magazine reported that the "invited audience" consisted of "200 scientists, photographers, motion-picture experts, and newspaper men." The writer described "a motion picture in which the characters did not appear to be on a flat screen, but seemed to be moving about in locations which had depth exactly like the real spots where the pictures were taken. This stereoscopic picture has been developed by a western inventor, and a six-reel dramatic production made entirely by his method has already been completed." The "western inventor" was Harry K. Fairall who built a dual-band stereoscopic camera with two lenses "set at a distance from each other equal to the average distance between the human eyes."[16] Fairall had worked for five years developing the camera, which contained "two mechanisms propelled by one crank shaft," and eventually filed over twenty patents covering different features of the camera. A photograph of Fairall's single-body, twin-strip stereoscopic motion picture camera appears with the *Popular Mechanics* article, and also in a February 1923 article in *Scientific American.*[17]

With U.S. patent no. 1,784,515 (December 9, 1930), titled "Binocular Nonstop Motion Picture Camera," Fairall provided a detailed description of "a superior advancing mechanism which is primarily for use in cameras, but with minor alterations may be advantageously used in projectors." The advancing mechanism was "continuous in motion," not intermittent, and made "practically no noise." Most important, the advancing mechanism was "so designed that it may comprise two co-acting advancers and therefore be used to advantage in stereoscopic apparatus." Fairall's patent included five pages of highly detailed drawings illustrating how "the frames of adjacent strips of film be paired so they are projected together just as they are exposed. If two frames not exposed at the same instant are shown on a screen, the image will be inaccurate." An additional innovation of Fairall's invention was "to provide a means for exposing the edge portions of corresponding frames of two films in order to mark them so that they may be easily and correctly paired."[18]

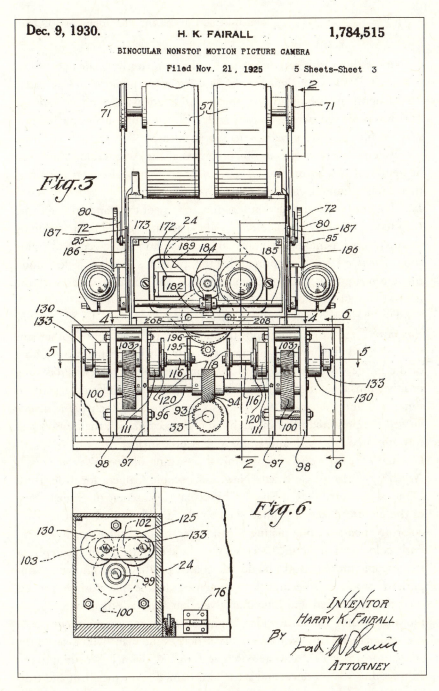

Dec. 9, 1930. H. K. FAIRALL 1,784,515

BINOCULAR NONSTOP MOTION PICTURE CAMERA

Filed Nov. 21, 1925 5 Sheets-Sheet 3

Fig.3

Fig.6

INVENTOR
HARRY K. FAIRALL
BY
ATTORNEY

Harry K. Fairall's dual 35-mm stereo camera, titled "Binocular Nonstop Motion Picture Camera," used to film *The Power of Love* (1930).

Interestingly, this patent, granted at the comparatively late date of December 9, 1930, was assigned to the Multicolor Corporation of California. This assignment is indicative of the two different ways that *The Power of Love* was projected stereoscopically and, indirectly, may shed some light on why the film never received a stereoscopic release.

When *The Power of Love* was projected at the Ambassador Theater in September 1922, both *Scientific American* and *Popular Mechanics* reported that it was exhibited with dual projectors. According to the article in *Scientific American:* "Two projection machines of any of the standard makes are required for showing the picture. . . . These are interlocked in operation by a simple attachment and two films are projected on the screen at the same spot." *Popular Mechanics* reported that the "two films" were projected "upon the screen simultaneously and superimposed one upon the other, with the use of two standard projection machines."[19]

For the stereo cinematography, as stated in the article in *Scientific American,* "one lens photographs through a green filter and the other lens through a red filter, thereby giving two negatives which contain everything of the image within the scope of the complementary colors." Processing positive films from the negatives, "the print from the red filtered negative is tinted red and that from the green is tinted green." When the audience viewed the projected films through red and green glasses, the motion picture appeared as a black-and-white stereoscopic image.[20]

In 1927, five years after the initial screening of *The Power of Love,* a *New York Times* article titled "New Stereoscopic Camera for True Third Dimension" noted that in Fairall's "binocular stereoscopic machine," "each of the pictures taken is printed on a single, double-coated positive. By a coloring apparatus, one picture is printed in red and the other in green." Early color motion picture processes were linked to the use of anaglyph for stereoscopic movies. Both the double-coated positive, also known as "duplitized" stock, and Technicolor's cemented double-film positive process used the orange/red and blue/cyan dyes that were effective for anaglyph motion pictures. Robert F. Elder, "a photographic and chemical engineer," was designated in the *New York Times* article as the man "who evolved the color process for this new stereoscopic film." William J. Worthington, a "former director, producer and actor" was named as the president and treasurer of the corporation controlling the patent rights to Fairall's binocular stereoscopic camera. Making single-strip anaglyphic projection of *The Power of Love* possible increased the film's chances for release.[21]

In the time between the first screening in Los Angeles of *The Power of Love* (1922) and the 1927 *New York Times* article, the Ives-Leventhal *Plastigrams* and *Stereoscopiks,* short anaglyphic novelty films, had been widely released on a national level. These short films had been shot by William T. Crespinel, a pioneer in motion picture color processes. Interviewed by his son William A. Crespinel in a 2000 issue of *Film History: An International Journal,* Crespinel recollected meeting Fairall shortly after moving to Hollywood in 1926. "After I had been on the coast about three months," Crespinel recalled, "I was approached at lunch one day by a man who introduced himself as Harry Farrell [*sic*]. He explained that he was with the Stereoscopic Binocular Film Company."[22] Fairall was familiar with Crespinel's work, was interested in hiring him, and asked him to meet Worthington in his office.

After meeting Worthington, Crespinel discovered that the Fairall 3-D system "turned out to be a duplicate of the basic system we had used. . . . What we had accomplished, apart from depth of the picture, and which was of interest to Worthington, was that we created the effect of images protruding from the screen. Could I obtain the same effect for him?" Crespinel, apparently unaware of the existence of *The Power of Love,* learned that Worthington planned to make a stereoscopic feature 6,000 feet in length and wished to incorporate Crespinel's "novelty method" for selected sequences. "Mr. Worthington," Crespinel replied, "I must tell you this. If you produce such a film, I doubt whether any exhibitor will show it in his theatre more than a few times."[23]

Crespinel explained:

> We had made six of these novelties and our experience had proved that the footage of each must be limited to between 400 and 500 feet. Even with that limitation, many of the audience became affected with eyestrain, dizziness, headache and nausea, the reason being that the eyes are subjected to an unnatural vision of (1) looking through complementary coloured glasses and (2) having the eyes focus on a distant object which is suddenly brought to a few inches of the eyes in a matter of a second or so. The screening time of our films was no more than five minutes. Thinking in terms of a film with a projection period of almost an hour just won't work.[24]

When informed by Worthington that the Stereoscopic Binocular Film Company had invested $50,000 in Fairall's process, Crespinel recommended producing "commercially saleable two-colour prints." Worthington and his associate Rowland V. Lee, a writer and director, were interested. In Crespinel's words "They recognized the need for practical colour films as against impractical 3D films and were prepared to abandon the old for the new."[25]

Fairall, Worthington, and Crespinel subsequently formed an association at the Multicolor Films Corporation in Los Angeles. Fairall's Binocular Nonstop Motion Picture Camera patent, first filed in 1925, was assigned to Multicolor when it was finally issued on December 8, 1930.

William Van Doren Kelley

William Van Doren Kelley and his efforts to create color cinematography as early as 1913 through a company named Panchromotion indirectly led to the first commercial exhibition of stereoscopic motion pictures in December 1922 in New York with anaglyphic movies he designated as "Plasticons." Kelley's Panchromotion additive process, using a rotating disk with red-orange, blue-green, blue-violet, and yellow, was short-lived. Kelley and Charles Raleigh had been granted a U.S. patent (no. 1,217,425) for the process in 1917.[26] This design suffered from the same defects as Kinemacolor and Friese-Greene's Biocolour process.

With a historical overview of color processes written in 1929 while he was a technical advisor at Multicolor Films, William Crespinel pointed out that all motion picture chromatic systems were either additive or subtractive. "The additive methods of color photography," he wrote, "are formed by the addition of complementary colors; the positive, itself, carries no visible color, only latent color values." But for the subtractive process, "the colors are visible on the film, itself, and these results are obtained by various means, such as mordants and dyes, high efficient chemical tones or by the imbibition or transfer process, and white is formed by the absence of complementary colors."[27]

Kelley was a prolific writer and filer of U.S. patents, as well as a color technologist. In a 1918 *Society of Motion Picture Engineers Journal* article titled "Natural Color Cinematography," Kelley assessed color systems then current for motion pictures. He pointed out that "Natural Color Motion Pictures" as opposed to "Color Motion Pictures" "may be understood as being those photographed so that the colors are selected entirely by optical and mechanical means and reproduced again in a like manner."[28]

One of the motion picture color systems that Kelley identified in his article was Douglass Color, a two-color additive system developed in 1916 by Leon F. Douglass of San Rafael, California. With U.S. patent no. 11,313,587 (August 19, 1919) titled simply "Cinematography," Douglass described a twin-lens dual-band motion picture camera which could "produce color and stereoscopic effects in motion pictures" by using complementary color screens, with each lens exposing left- and right-eye views alternately and leaving every other film frame blank. A double frame pull-down was used to advance the film. With the two lenses horizontally displaced, this system could be used for photography solely in color or for stereoscopic cinematography. A single positive film was printed from both negatives, with the images appearing in succession. The print was then projected "by a single lens projecting machine employing a shutter having a red window and a green window." On February 14, 1918, in New York City, Douglass presented a color demonstration of his cinematography with a program depicting bathing beauties and scenic views of Yosemite and Yellowstone National Park. Douglass was subsequently granted a U.S. patent on September 19, 1922 (no. 1,429,495) describing a single-lens motion picture camera with a "prism block" to be used with a shutter mechanism for stereoscopic cinematography.[29]

In 1917, Kelley began to have real success with his Prizma color process, which used double complementary pairs of images to take photographs. With the filming running at the standard sixteen frames per second, one pair recorded red-orange and blue-green colors and the other orange and blue. A short film of 1917, *Our Navy,* was released in the additive Prizma color process.

By 1919, Kelley had developed the subtractive Prizma color process with orange/red and blue/green printed on duplitized film, a double-coated positive, from two panchromatic negatives. In January 1919, Kelley began making weekly releases of short color films that could be projected on a standard projector. Kelley was a friend of Hugo Riesenfeld, who would prove to be an important influence for exhibition of anaglyphic novelty films in the 1920s. Riesenfeld was the directing manager of the Rivoli Theater and Roxy Theater chain. Many Prizma color films premiered through the Roxy chain, including *Bali: The Unknown,* the first Prizma color feature, in 1921 and the Vitagraph epic, *The Glorious Adventure,* starring Diana Manners and Victor McLaglen, in 1922.

It was a definite step forward that prints made for Prizma color and Plasticon movies were double-coated positives on a single strip of film with orange-red on one side and blue-green on the other. This single-strip two-color process was still used after the Prizma color company folded and subsequently became the Multicolor Company of Hollywood in 1929.

In 1918, Technicolor was working on its two-color dye imbibition process transferring color to film in the laboratory. Technicolor used two positives cemented together, base to base, with the emulsions of the complementary colors facing out. The disadvantages were obvious. It was impossible to achieve sharp focus on both sides of the film at once. And the prints received considerable scratches to the emulsion over long periods of exhibition. Another problem was "cupping" of the film that would curl as a result of the double-sided emulsions and cemented positives.

Plasticons

Kelley's Plasticon film was given its public premiere during Christmas week of 1922 at Riesenfeld's Rivoli Theater in New York with the novelty short *Movies of the Future,* depicting scenes in and about New York City. In an October 1923 article in the *Society of Motion Picture Engineers Journal* titled "Stereoscopic Pictures," Kelley discussed the Plasticon movies and characterized them as "simply an adoption of the Du Hauron Anaglyph. The system was adopted because of its simplicity. The camera pulls down two picture areas at a time and exposes both at once. Each lens is provided with a prism so that one lens sees about 1⅜" to the right of center, while the other sees the same distance to the left." The overall interocular distance, then, would be 2¾ inches. Kelley also noted that "the same camera that is used for color pictures was utilized by making a new mounting for the prisms and lenses interchangeable with the color lenses."[30]

Kelley cited a press notice from the *New York Times* about *Movies of the Future* in his article: "These stereoscopic pictures have to be seen to be appreciated. Persons accustomed to viewing ordinary flat pictures cannot easily anticipate the effect of seeing every object in a scene stand out with length, breadth and thickness as it does in real life. They do just that in Mr. Kelley's short film."[31]

In addition to numerous patents for color processes, Kelley, along with Dominick Tronolone, was granted a U.S. patent (no. 1,729,617) in 1929 for "a means and method of exhibiting to an observer a simulation of motion as an image, on a screen which will produce a stereoscopic effect when properly viewed." The patent, titled "Stereoscopic Picture," described a

STEREOSCOPIC PICTURE
Filed July 24, 1924   3 Sheets-Sheet 1

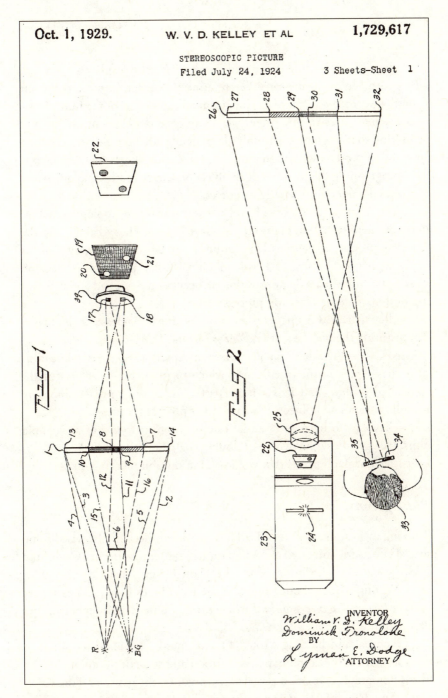

William V. D. Kelley's "Stereoscopic Picture," a curious patent addressing
production of shadowgraphs (1929).

single-lens camera with a prism and two horizontally displaced apertures, along with "color means for differentiating the image formed to the left from that formed to right of the longitudinal center line of the film." Curiously, Kelley's stereoscopic patent went into great detail about filming and exhibiting shadowgraphs. The only claim finally cited on the patent is for a "method of producing motion pictures in stereoscopic relief which consists in photographing the shadows of an object on a curtain, the shadows being overlapped colors from two light sources."[32]

Kelley's article gave a brief history of stereoscopic technologies, including those derived by Du Hauron, Theodore Brown, Harry Fairall, and the Teleview process of Laurens Hammond. After closing the reading of his paper with a discussion of Plasticon films, Kelley passed out anaglyph glasses and projected a demonstration of "scenes made in and about Washington D.C." in the Plasticon process.[33]

Kelley delivered his paper some months after the release of the Plasticon pictures. The Ives-Leventhal *Plastigrams* and *Stereoscopiks* were yet to be released. "It is difficult to say just where stereoscopic pictures will lead us since they represent the true form in which pictures should be presented," concluded Kelley, "and it is a more perfect method of photographically recording scenes and objects than any other process yet presented."[34]

Kelley was to have an essential role in the next round of anaglyphic films. He would use the two-color double-coated process to make the Ives-Leventhal anaglyph prints that would run in motion picture houses.

Plastigrams

Working as a cameraman for Max Fleischer, Jacob Leventhal devised some special effects for animated cartoons that led to their use in the "Out of the Inkwell" cartoons featuring Koko the Clown. Leventhal, together with John Norling, wrote about some of these techniques in a 1926 *Society of Motion Picture Engineers Journal* article titled "Some Developments in the Production of Animated Drawings."

Leventhal and Norling wrote: "One of the most interesting of recent developments is the combination of cartoon and straight photography. . . . The [photographic] image is projected down upon the paper, and the drawings are made to fit with action of the straight photography characters. Two negatives are made of the cartoon and double printed to secure the desired effect."[35] Such techniques would give the Fleischer brothers' cartoons, even in 2-D, a very stereoscopic feeling.[36]

Crespinel recollected first meeting Leventhal through Max Fleischer. Leventhal had just invented the "bouncing ball" concept for sing-along cartoons, which audiences loved. Crespinal recalled:

> I would play cards with Max Fleischer and his staff. Among the players was Jack Leventhal, a friend of the Fleischer family. . . . One day Jack told me he had an idea. He planned to make 3D films based on the anaglyph principle of viewing red and blue selectively coloured stereoscopic pictures through complementary filters. I accepted Jack's invitation to join him in the experiment.[37]

Two initial requirements had to be met: (1) a motion picture camera with two lenses and (2) a way of creating colored prints. Leventhal and Crespinel visited Frederic Ives in Philadephia who placed two cameras side by side, which provided optical centers for the lenses of 2 ¾ inches, perfect for stereoscopic photography. The two cameras were hinged together with a common drive shaft to work as a unit. Ives had Leventhal and Norling take the stereo camera to make tests.

For the anaglyphic motion picture prints, Leventhal and Crespinel contacted William Van Doren Kelley who selected suitable dyes to use with double-coated film and lorgnette red/blue anaglyph viewing glasses obtained from Freedman Cut-Outs in New York. Leventhal and Crespinel began stereoscopic filming. "We made a sample film in and around Washington, D.C.," recalled Crespinel. "Jack showed the film to Dr. Riesenfeld, managing director of the Rialto theatre in New York, who enjoyed the 3D effect, but said it lacked novelty. 'Show me something like Ziegfeld has at the "Follies" and I might be interested.'"[38]

Hammond's shadowgraph produced an exaggerated 3-D effect that was difficult to achieve with two camera lenses in parallel alignment. "The novelty was the startling illusion of objects appearing to come from the screen into one's face," explained Crespinel. Leventhal and Norling went back to Ives in Philadelphia, who subsequently supplied them with "a pair of odd-shaped prisms to be placed in front of the lenses." The prisms converged the optical axes of the two camera lenses, thus producing negative parallax with off-the-screen imagery. It permitted "an object to come within a certain distance of the lenses," in Crespinel's words, and it "reacted in a similar way in the projected image."[39]

Crespinel and Norling filmed vignettes that broke the "fourth wall" of the motion picture screen, foregrounding spectatorship and display, acknowledging the viewer with a visual shock. Riesenfeld wanted Crespinel and Leventhal to produce a cinema of attractions, and they delivered it. These were stereoscopic visual novelties that violated the audience space. "We made a film using the new technique," said Crespinel, "a baseball thrown into one's face; a water hose turned full on; a spear aimed at one, etc. At a private showing for Dr. Riesenfeld, he said, in effect, 'Great. I'll book it.' There had never been anything like it on the screen before."[40]

In point of historical fact, there *had* been something like it before, as seen when Edwin S. Porter's bandit raised his gun and fired at the audience in *The Great Train Robbery.* Now, two decades later, Porter's thrilling acknowledgement of the audience had been heightened in a stereoscopic display. By converging the optical axes of the lenses in the cameras, the stereoscopic imagery was made to come out into the audience space. Crespinel and Leventhal wisely eliminated backgrounds in the 3-D vignettes to minimize eyestrain for the audience.

Plastigrams played in exclusive engagements in New York at Riesenfeld's Rialto and Rivoli Theaters before general release. Then, in the January 23, 1924, issue of *Motion Picture News,* Educational Film Exchanges announced that *Plastigrams* would "soon be available to every exhibitor in the United States in the form of an Educational Pictures Special."[41] The headline of the story boldly exclaimed, "'Third Dimension' Effect Realized, Educational Announces Long Sought Motion Picture Improvement Is Attained." The Educational Film Exchange specialized in short films of all kinds. It made twice weekly releases of "Kinograms, The Visual News of All the World," presented as song-pictures, as well as one-reel and two-reel comedies and documentaries. Educational Pictures declared through the trade magazines that March 1924, the month they released *Plastigrams,* was "the largest month program of its season."[42]

Earl W. Hammons, president of Educational Films Corp., was a proponent of the short film. "Why Not a Short Subject Theatre?" Hammons proposed in an article in *Motion Picture News.* "Why not have a Short Subjects theatre—a sort of informal 'drop-in' kind of theatre—in the heart of every big city, where those who like variety can always count on finding it?" wrote Hammons.[43] He was no less vocal in promoting *Plastigrams:*

I believe "Plastigrams" will mark the greatest forward step the motion picture industry has ever seen. It is a revolutionary invention, the possibilities of which cannot be estimated at the present time. It introduces absolutely new sensations to the observer by making him seemingly present at the scene on the screen.

Actors approach the observer within arm's reach; the spectator feels that he can reach out and touch the artists. It is much more than a picture in relief, as actors and articles approach the spectator in such lifelike manner that audiences instinctively recoil when, for instance, a stream of water is directed at them from the screen.[44]

The January 23 article announcing *Plastigrams* also credited "Frederic Eugene Ives, inventor of the half-tone photoengraving process, the color plate process used at the present time as well as many inventions" for stereoscopic photography and "Mr. Leventhal, his associate [who] has been identified with the technical and experimental side of the motion picture industry for many years."[45]

March 23 was announced as the release date for *Plastigrams,* and, in the March 15 issue, a full-page ad announced "AT LAST! THE THIRD DIMENSION MOVIE AVAILABLE NATIONALLY FOR THE FIRST TIME" and accompanied a line-drawn image of a man in a hat pointing directly at the reader. The following advertisement, quoting *Billboard* magazine, is illustrative of Educational Pictures' approach to promoting the novelty of *Plastigrams:*

If you want to give your patrons a treat, and incidentally show a novelty that will create a lot of talk, get hold of the short reel of Third-Dimensional pictures called Plastigrams * * * Shown at the Rivoli last week * * * Some amazing effects were obtained * * * Shouts of amazement and surprised laughter from the unsuspecting audience * * * Plastigrams are a great novelty

Further hyperbole in the ad proclaimed that *Plastigrams* marked "one of the greatest feats of motion picture production ever achieved, this great novelty will bring shrieks of laughter and excitement and set your patrons to talking. Ask to see it at your exchange."[46]

The promotional treatment that Educational Pictures gave to *Plastigrams* paid off. *Motion Picture News* reported that it was "widely booked" with first-run houses "keenly interested."[47] At the end of March, when the

Saxe chain opened a new $2 million theater, the Wisconsin in Milwaukee, every short subject on the opening week's bill was from Educational Pictures. *Plastigrams* was part of the program on opening day. The majority of the Paramount theaters, comprising 130 houses, had contracted for the film, and representative theaters—in addition to the Rivoli and Rialto in New York—included the Rowland and Clark circuit of Pittsburgh with twelve theaters; the A. H. Blank circuit in Omaha and Des Moines with eleven houses; and the Capitol, Adams, and Madison of the Kunsky theaters in Detroit.

Stereoscopiks

With the success of *Plastigrams,* there was a demand for more stereoscopic novelty films. In a February 1925 issue of *Motion Picture News,* Pathe Exchange announced distribution of the Ives-Leventhal *Stereoscopiks,* "a series of 'Third Dimension' single reel subjects." Again the novelty appeal was highlighted: "It is said they will make the action taking place on the screen appear as if actually being projected out over the footlights into the midst of the spectators."[48]

March 22 was given as the release date for the first *Stereoscopiks* film, *Zowie. Lunacy,* the second in the series was scheduled for release on May 17. William Crespinel recalled the filming of *Lunacy:*

> Then we made one at Luna Park, the great amusement centre at Coney Island. We titled it Lunacy which is just about what it was. Every scene was a natural to evoke screams from the audience. The roller coaster, then the tallest in the world, was a wow. "Shoot the Shoots," smack bang into the water! The revolving Ferris Wheel, etc. After a showing, Dr. Riesenfeld booked it immediately.[49]

Lee DeForest began producing his Phonofilms, with sound as an optical track on the film, in 1922. R. M. Hayes noted that "*Lun-acy* [*sic*], like the reissued Plastigrams, was supplied with a Photofilm [*sic*] soundtrack of music and possibly effects for selected engagements, though generally exhibited as a silent film."[50] Once again, Hugo Riesenfeld was to prove instrumental in this cinematic innovation. When DeForest introduced Phonofilms, he had great difficulty exhibiting them. His biographer, Maurice Zouary, noted:

Most of the theaters in the country were controlled by the producing studios which had no desire, of course, to promote a competing product. New York City's Rivoli was the exception, however, and continued to serve as the premiere showcase for Phonofilm productions, week after week. This arrangement was due solely to the faith and the foresight of Hugo Reisenfeld [*sic*], and not to any nationwide policy of the chain to which his theater belonged.[51]

When Leventhal's bouncing ball concept was introduced by the Fleischer brothers in the sing-along cartoon, Hugo Riesenfeld was well aware of the possibilities. Zouary explained:

> Seeing a natural possibility for a sound-synchronized subject, Riesenfeld brought the Fleischers and DeForest together, to work out a sound cartoon of the "Sing-a-Long." As a result of their initial collaboration, animated song cartoons became more elaborate and effective. With the sound emanating directly from the screen in full synchronization, the songs came alive, almost compelling the audience to sing along. Sometimes a cartoon character such as Koko the Clown ("Out of the Inkwell") introduced the songs or bounced on every syllable.[52]

With his connection to the Fleischer brothers and Hugo Riesenfeld, it was a natural development that Leventhal would incorporate sound into both *Plastigrams* and the *Stereoscopiks* films.

Release of two more *Stereoscopiks* films continued with *Ouch!* and *Runaway Taxi,* both on December 17, 1925. Pathe gave the *Stereoscopiks* a big build-up, with full-page ads in *Motion Picture News* heralding "A Series of Four Short Reels, the Most Novel of the Novel." One full-page ad depicted a woman swimmer emerging with a splash of water off the screen: "A girl swims toward you. She seems to leave the screen and come right out over the heads of the audience into your very eyes." Again, the success of the films at Riesenfeld's Rialto was cited: "Are you surprised that audiences at New York's Rialto screamed and laughed and applauded?" Another ad with a line drawing of a woman swinging out into the audience declared, "At New York's Rialto audiences have been swept with waves of laughter as they watched the super-novel Stereoscopiks on the Screen. Short in length, long in interest, big in audience values."[53]

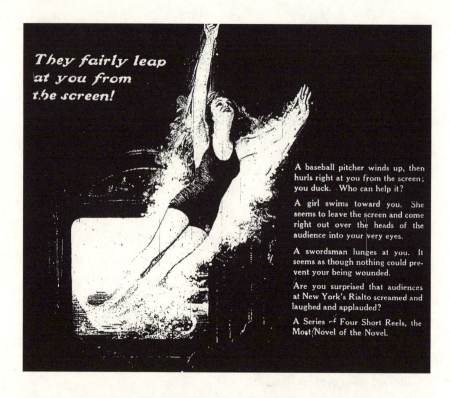

They fairly leap at you from the screen!

A baseball pitcher winds up, then hurls right at you from the screen; you duck. Who can help it?

A girl swims toward you. She seems to leave the screen and come right out over the heads of the audience into your very eyes.

A swordsman lunges at you. It seems as though nothing could prevent your being wounded.

Are you surprised that audiences at New York's Rialto screamed and laughed and applauded?

A Series of Four Short Reels, the Most Novel of the Novel.

© 1925, A. C. CO.   66   A Modern Mermaid

Detail from *Stereoscopiks* ad *(above)* (1925), matched by an American Colortype litho stereoview card *(below)* (1925).

To hedge their bets and to make sure to please Hugo Riesenfeld, Crespinel and Leventhal included shadowgram-style vignettes in their stereoscopic novelties, depicting silhouettes of a fisherman casting a line at the audience, men erecting planks out into the audience space, or a baseball game with a runner sliding directly at the spectator. These motifs were also repeated with close-up direct photography depicting a Charlie Chaplin character hurling a pie at the audience, a dandy with a sword lunging out, and a baseball player throwing a ball directly at the viewer.[54]

In 1925, the American Colortype Company issued a novelty set of lithographed stereoview cards that directly illustrated the very same motifs, from the swordsman to Charlie Chaplin, from the girl swimmer to the baseball motif. The stereoviews are copyright 1925 to "A. C. Co.," feature black backgrounds, and use a hyperstereo effect with imagery coming right out at the viewer, just as with the *Plastigrams* and *Stereoscopiks* films. The actors in the stereoview cards even bear a striking resemblance to those in the films. Interestingly, one of the A. C. stereoview cards, depicting a ballerina extending a foot out at the viewer, bears the imprint "Courtesy of Pathe Exchange, Inc." suggesting a licensing arrangement of some kind between Pathe and American Colortype.

After producing six short anaglyphic novelty films with Ives and Crespinel, Leventhal published a somewhat gloomy article in the November 1926 issue of the *Society of Motion Picture Engineers Journal,* titled "The First Use of Stereoscopic Pictures in Motion Picture Theaters." The article was a three-page summary of the Teleview and anaglyph processes as they had recently been used in motion pictures. "In the field of stereoscopic reproduction in motion pictures there seems to be very little if any hope of development," opined Leventhal. "The obstacles seem insurmountable." In Leventhal's view, the primary problem was that the "aid of special apparatus" was necessary for stereoscopic viewing. Characterizing Teleview and its effects as "excellent," he noted that "there are difficulties, as you may imagine, on the practical side."[55]

After a brief description of the anaglyph process, Leventhal asserted that the biggest difficulty in exhibition was "the handling of the spectacles in conjunction with the film." He noted, "It was obvious at the beginning that if the exhibitors were to accept this kind of picture, it would be necessary to emphasize the spectacular side and make scenes that would startle the audience, rather than views of streets and scenery." Curiously, Leventhal did not address the technical aspects of stereoscopic filming in the ar-

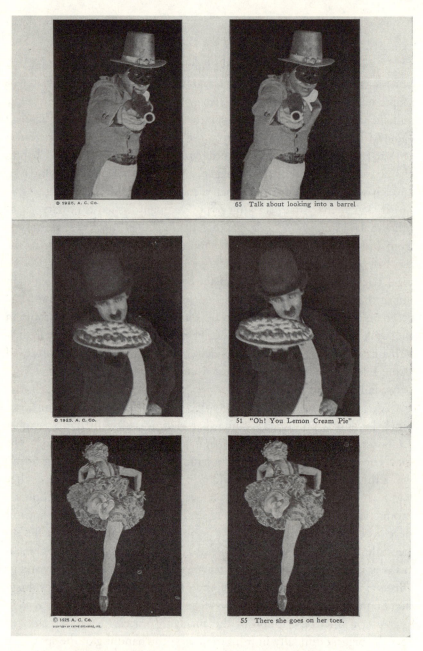

The American Colortype litho stereoviews featured motifs identical to the *Stereoscopiks* films; notice the attribution to "Pathe Exchange" at the bottom of the ballerina card (1925).

ticle but, instead, enumerated the extra costs of the glasses and color print stock (eight times as much as black and white), as well as the difficulties the "porter" experienced when cleaning up the theater after the show: "The spectacles cannot be inhaled by his vacuum cleaner and must be picked up by means of physical exertion. This annoys a great many porters." Leventhal noted that in one or two instances the exhibitor neglected to distribute spectacles, thereby displeasing the audience. Because a viewing apparatus was necessary, Leventhal wrote, a stereoscopic film "can never occupy more than a few minutes on a program."[56]

After reading his paper to the Society of Motion Picture Engineers members, Leventhal presented a film that "was a modification of the usual anaglyph stereoscopic film, the red image depicting a scene entirely different from that given by the green image" with the audience having the choice of viewing either "a happy or a sad ending to a picture by looking through either the red or green filters."[57]

This film was undoubtedly the last of the six anaglyphic shorts that Leventhal and Crespinel filmed. Crespinel gave an accounting of the film, which was titled *As You Like It* with the subtitle *Not Shakespeare*. The picture begins in straight black and white. A young wife waits for her husband to come for lunch from the sawmill. He is late. A villain at the sawmill blackjacks the young husband, lays him on a board, and starts the saw blade. The wife jumps in her car, arrives at the sawmill and enters. A title card informs the audience to "PUT ON YOUR GLASSES." In Crespinel's words:

> If one looked with the left eye (blue) here's what they saw: Exterior of cottage—group of people standing around—front door opens—six men appear carrying a coffin on their shoulders. As they approach camera, the coffin separates lengthwise, down the middle, and is carried by three men—the villain's scheme was successful.
>
> Through the right eye (red) we see wife enter mill room—sees, in horror, her husband's predicament—fumbles around to locate turn-off switch—and then the embrace.[58]

Leventhal may have had a pessimistic outlook regarding the future of the stereoscopic film. He, nevertheless, continued working in the field, and, with John Norling, made further important contributions to the genre.

8

Essaying Utopia

A Few Stereoscopic Patents

While the motion picture industry consolidated in the 1920s with greater technological developments for color and sound, utopian inventors continued to file patents for three-dimensional moving images. These efforts were frequently attempts to simplify production of stereoscopic movies and generally did not come to fruition. Undoubtedly, it was the growth of motion pictures as entertainment that spurred attempts at stereographic innovation, which continued to present itself as the ultimate form of crowning realism for the moving image.

A U.S. patent (no. 1,363,249) of 1920, granted to Fred N. Hallett of Seattle, Washington, for a "Moving-Picture Camera," described an improvement "to provide a camera that will take pictures having a stereoscopic effect similar to that produced in the human eye." The camera was designed with "two lenses, the light from which is reflected on the film at a single point, the rays from each lens cooperating to produce the desired effect." Hallett's simple design used mirrors positioned behind the twin lenses. "By means of the two lenses and the mirrors, the images passing through each lens [sic] is reflected on the single film and combined, as in the human eye, to produce the same stereoscopic effect."[1] The internal mirrors were adjustable and placed behind the twin converging lenses to reflect directly on to the single strip of film. Hallett's patent, though granted, may not have had sufficient detail for reduction to practice. It is also likely that a separate, and dedicated, system for projection of the single-strip stereo pairs of images would have been required. This patent did foreshadow, however, numerous subsequent efforts, from the 1930s up to the 1980s, to build a single-strip 35-mm stereoscopic camera.

A 1923 U.S. patent (no. 1,477,541) assigned to Clement A. Clement by Axel Bors-Koefoed of Houston, Texas, titled "Motion Picture Machine" describes and pictures two separate stereoscopic projectors for single-strip 3-D films. Two separate pages of figures depicted 35-mm film with side-by-side stereo pairs in a single frame of film, as well as a single strip of 35-mm film with vertically alternating left-right images. The object of Bors-Koefoed's invention was "to provide a machine . . . whereby images taken at optic angles may be simultaneously projected on the screen so as to give a relief effect."[2] Bors-Koefoed's patent includes provision for a stereoscopic motion picture camera and rotating disk shutters used in front of both projectors. The patent, which does not go into great detail, seemed to be an attempt to protect two different single-strip stereoscopic film configurations.

A 1921 U.S. patent (no. 1,396,651) granted to Edgar B. Moore for a "Stereoscopic-Motion-Picture Mechanism" was virtually identical to Hammond's Teleview patent, albeit slightly less detailed. Moore's patent for "a new and useful Improvement in Stereoscopic-Motion-Picture Mechanisms" described "a mechanism whereby a pair of stereoscopic pictures, or two stereoscopic series of motion pictures, can be projected on a screen and the binocular effect of depth and solidity observed by all in the audience." Moore's mechanism, like Hammond's, projected left and right images on the screen in rapid succession using a rotary shutter that was run by a synchronous electric motor. An improvement was claimed with the novelty of Moore's invention for the combination of rotary shutters and electric motors, "whereby exact synchronism is preserved at a high speed thus giving an improved stereoscopic effect."[3] Moore neglected to specify the exact rate of the high speed, however, thus failing to sufficiently differentiate his patent from Hammond's.

Motional Perspective

Inventors have perennially attempted to create a stereoscopic process that does not require a viewing apparatus. The technique of employing camera motion to produce horizontal parallax was attempted by several inventors, whose aim it was to achieve an autostereoscopic movie. This technique had some singular drawbacks and, when presented as an oscillating image, simply did not work. That, however, did not stop the inventors in their endeavors to realize this most unique variation of the utopian dream.

With a 1920 U.S. patent (no. 1,351,508) titled "Method of Producing Stereoscopic Effects in Motion Pictures," Raymond A. Duhem and John D. Grant described an invention in which the lens of a single motion picture camera was "shifted from side to side with its principal optical axis directed toward the center of the object or scene being photographed." Successive exposures of the object were to be taken from different positions that approximated the left- and right-eye viewpoints of a stereo pair of images. Duhem and Grant claimed that the dissimilar images were combined by persistence of vision when rapidly projected on the screen. They asserted that "the chief novelty of the present invention" consisted of "a centering of the principal optical axis of the camera on a fixed point in all positions of the camera."[4] The result of this cinematography was that certain parts of the image would jitter or "stutter" horizontally during viewing. The authors made drawings showing the single camera lens converged from different positions upon a fixed point and claimed that this would eliminate or minimize the apparent movement in the images. At no time was binocular stereopsis affected, and both eyes of the viewer saw identical, stuttering images.

Waldon S. Ball made use of camera movement in a 1920 U.S. patent (no. 1,351,502) for a "Method of and Apparatus for Taking Moving Pictures." Ball claimed his invention had "for its object to obtain new and improved results in the photographing of objects with the subjects appearing in relief." Ball's patent illustrated a camera suspended on a wire strung between two trees. The camera moved horizontally during filming. Ball asserted that his apparatus would produce "proper timing of the travel of the camera and feed of the film." Ball wrote that the results obtained "cannot be called true stereoscopic effects, but provide pictures in which the subjects appear in relief without being blurred or distorted and thus cause them to have a very, life-like appearance."[5]

These patents made use of what British pioneer Theodore Brown had called "Motional Perspective." Similar in style to the tracking shots of the novelty era, motion parallax produced a feeling of stereoscopic depth using monocular cues. Cecil Hepworth had used the technique in 1905 for a series of short films that he entitled *Stereo Scenics*.[6] After observing the stereoscopic effects in camera tracking shots, Brown devised a means of horizontally oscillating a small subject before the motion picture camera. Brown

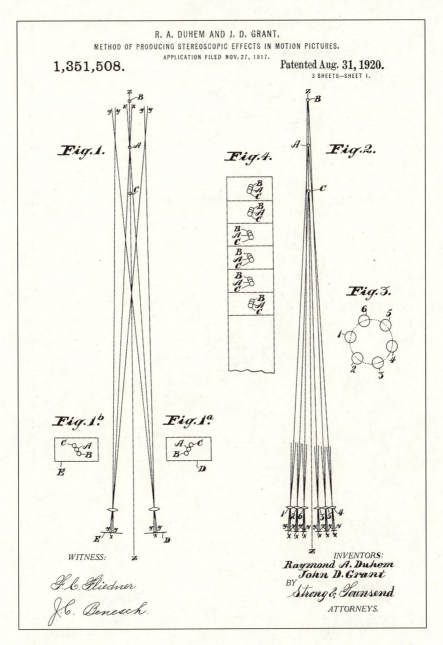

R. A. DUHEM AND J. D. GRANT.
METHOD OF PRODUCING STEREOSCOPIC EFFECTS IN MOTION PICTURES.
APPLICATION FILED NOV. 27, 1917.

1,351,508.

Patented Aug. 31, 1920.
3 SHEETS—SHEET 1.

Fig.1.

Fig.4.

Fig.2.

Fig.3.

Fig.1.ᵇ

Fig.1.ᵃ

WITNESS:

F. C. Fliedner

J. C. Benesch.

INVENTORS:
Raymond A. Duhem
John D. Grant
BY
Strong & Townsend
ATTORNEYS.

Raymond A. Duhem and John D. Grant's "Method of Producing Stereoscopic Effects in Motion Pictures," an attempt at stereo through the use of motion (1920).

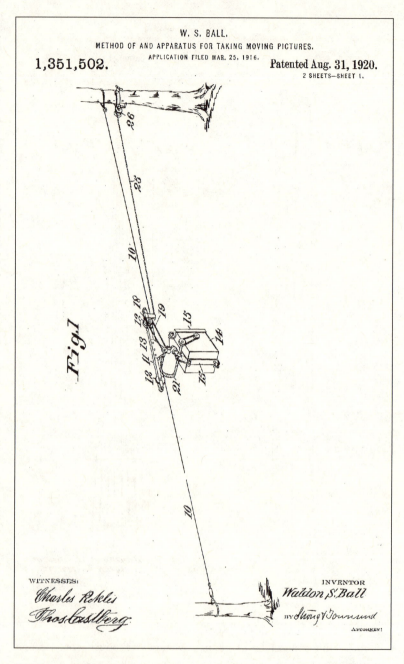

Waldon S. Ball's highly imaginative patent, "Method of and Apparatus for Taking Moving Pictures," used motion parallax for stereo (1920).

then devised an oscillating camera and wrote that "we may carry to the mind through a single eye impressions of binocular vision."[7] Tirelessly promoting this form of monocular stereo and despite discouragement from his peers, who were less than enthusiastic about the visual effects, Brown continued to work with motional perspective techniques for many years.

Spoor-Berggren—Natural Vision

One of the most drawn-out sagas of frustrated invention is that of George K. Spoor and Dr. Paul John Berggren and their attempts to create a stereoscopic motion picture system they called "Natural Vision." Spoor was the "S" and "Broncho Billy" Anderson was the "A" in the Essanay Film Manufacturing Company of Chicago. Spoor began working with Berggren in 1916, and early in 1921 they announced a three-dimensional camera with two lenses.

In January 1921, a headline in *Motion Picture News* exclaimed, "Three Dimension Photography Is Perfected." According to the unidentified reporter: "The importance of this announcement to the motion picture industry is realized when it is understood that this means that pictures on the screen can be given the third dimension . . . that is depth, making persons or objects on the screen stand out as if we were actually viewing them with our eyes in real life." Plans for commercial exploitation of the new "two-eye camera" had not been completed at the time of the announcement. Spoor also claimed that conventional cameras in use at the time could be made into stereoscopic cameras by the addition of a small attachment and that the regular projectors in use at the time could be fitted with a similar attachment for stereoscopic projection. The reporter compared stereoscopic and conventional "flat" movies by writing that "the difference may be explained by saying that the two-eye camera photography is like looking at a statue with a natural background, whereas the present picture is like a painting of the statue." Spoor was no stranger to hyperbole. He claimed that the large fortune he had earned as a pioneer in the motion picture industry funded his stereoscopic efforts despite "the enormous expense entailed."[8]

Spoor and Berggren had evidently modified the Natural Vision format for a screening on August 20, 1923. Edward Connor wrote:

> It required two large screens, 60 by 34 feet, onto which two pictures of the same scene were projected. They were taken with a two-lensed camera—the lenses were three inches apart—and pro-

jected through a nine pound lens containing two refracting elements. One of the screens was made of black threads and was translucent; the screen behind it was a solid with a reflecting surface. No glasses were required. Seven years' work and $1,000,000 had been expended on the new system.[9]

In April 1925, a notice in *Motion Picture News* announced that, "after eight years of preparatory work and expenditure of more than a million dollars," work on sets to be used in filming the first Natural Vision feature, *The Price of the Prairie*—a drama of the old west to be directed by Norman McDonald—had begun. It stated: "The camera in addition to providing stereoscopic effects allows the use of a screen of tremendous size" and that specially built projectors and new screens would be necessary to show the picture.[10] Yet despite the announcement, *The Price of the Prairie* was not subsequently released in Natural Vision.

Spoor continued to issue periodic notices to the press about Natural Vision. Then, in November 1929, *Danger Lights,* a feature-length railroad melodrama, purporting to be in Natural Vision and made at RKO studios, opened at the RKO State-Lake Theater in Chicago. Starring Louis Wolheim, Jean Arthur, and Robert Armstrong and directed by George Seitz, *Danger Lights* was projected with a wide-screen system called Magniscope and played to sold-out audiences at $1 a ticket. Spoor promoted the film as "the first picture to be made with this new process," and it played for seven weeks at the Mayfair Theater in New York.[11]

Several short films were shot for Natural Vision at RKO in 1930, including *Campus Sweethearts* with Rudy Vallee and Ginger Rogers. Karl Struss, Academy Award–winning cameraman, worked on some of these films at RKO and later stated that "they were not stereoscopic."[12] Years after its release, Spoor also acknowledged that *Danger Lights* had not been released in stereo. He claimed that he had been forced by RKO, under pressure from the Motion Picture Producers Association, to release the film in conventional 35-mm format. Ironically, Spoor passed away in 1953, just as a boom of stereoscopic film production was taking place in Hollywood.

Parade of Inventors

When George K. Spoor announced his perfected stereoscopic motion picture camera in 1921, an article in the *New York Times* noted that the announcement "attracted into the open reports of other and different machines for making pictures with three dimensions." One of these machines was invented by Charles V. Henkel and used a single lens that did not require special film or photography. Dubbing his process "Spectra," Henkel stated that it was "color value," as distinguished from actual color, which made his process work. "Color value exists in all positive motion picture prints, to a certain extent not visible to the eye as now projected," Henkel explained.[13] His process purported to make color value visible in giving an impression of depth on the motion picture screen.

Henkel had projected demonstrations of Spectra to his family and friends. He pointed out that "a one-eyed man sees things in three dimensions as well as a man with two eyes" and that "'solidity' existed in every picture, but could not be perceived unless emphasized by his process."[14] However, nothing further was subsequently heard about Henkel's process.

Under the heading of "3-D systems that were actually tested, not merely talked about, before they disappeared," Edward Connor has identified a few experiments of the 1920s.[15] Connor noted that a May 1923 screening on a single strip of film that required no glasses for 3-D viewing was made by the American Releasing Corporation in a rented projection room in New York City. Walter Parkes of Oakland, California, shot the scenes of the Grand Canyon that were projected and, according to the *New York Times,* demonstrated "hardly any distortion, and none in objects near to, it has a distinct advantage over other attempts."[16]

Another autostereoscopic process was promoted and described by D. Daponte, a Romanian inventor working in Paris, whose explanation of his system derived from twin-lens filming was quoted in the *New York Times:* "By rendering the lighting of each picture different and continuously varying it so as to throw a continuous equal constant stream of light onto the screen . . . the effect produced is a third dimension with full relief and depth visible to the naked eye."[17] Daponte also stated that he had demonstrated his system in London and promised that a feature-length film using the process was forthcoming.

Keating Randall, an inventor from Texas, announced his 3-D process in July 1929 after arriving in New York City to promote it to the Vitaphone Company. Randall's system was highly complex:

> The screen would be built up of sets of parallel planes whose thousands of lines would cross one another on the diagonal. . . . The Projector would be equipped with two lenses: one would project

images with ultra-violet rays, the other with light as deficient as possible in actinic rays. These two sets of images would be built up on the screen, producing a third dimension.[18]

A twin-lens mirror stereo attachment for a single camera, sometimes called a "beam splitter," was offered for sale in August 1929 by the British Filmcraft Company. The device produced a stereo pair, separated by a hairline, on a single strip of film. The asking price for the stereo attachment was $700. Similarly, an inventor named Donovan Foster described his Stereolite system in the late 1920s. Foster explained that the effect of depth with the system was created by twin beams of light captured, slightly offset, on a single strip of film.

Autostereoscopic systems, dispensing with binocular stereopsis, were frequently hopeful, but naïve, attempts to eliminate the need for a viewing apparatus. The numerous announcements in the 1920s of new stereoscopic systems for motion pictures, many unbuilt and untested, are evidence of both the widespread desire and the difficulty of achieving the third dimension in cinema.

The Case of William Waddell

With the tantalizing glimpses of the motion picture career of William E. Waddell, revealed to us in only a few publications, we see in faint outline the employment trajectory of a motion picture engineer who worked to give greater realism to the moving image, first with stereoscopy, then with sound in the 1920s, and, eventually, with large-format film.

It was Waddell who worked with Edwin S. Porter in creating the system used to project anaglyphic segments of *Jim the Penman* at the Astor Theater on the morning of June 10, 1915. The June 26 *Moving Picture World* review of the stereoscopic screening of *Jim the Penman* noted that Waddell was "formerly of the Edison Company."[19] According to Maurice Zouary, Waddell was "a capable engineer who had worked with the Edison Kinetoscope operation."[20]

In 1925, Zouary also reported, Lee DeForest had begun locating maverick motion picture theater managers who were willing to wire their theaters for Phonofilm sound and make adjustments for the new speed and aperture requirements. When DeForest had thirty-four theaters adding Phonofilm programs to their regular features, Waddell was hired to secure new acts for the Phonofilm shorts. Zouary wrote:

DeForest gave him [Waddell] the job of selecting acts from the dramatic and musical stage, opera, and vaudeville. Waddell was obliged to persuade the performers to work for little or no compensation in exchange for the possibility of immortalizing themselves in the new medium of talking pictures. In the months that followed, Waddell succeeded admirably in obtaining top-rank performers, indeed giving them a small measure of immortality.[21]

Initially, the DeForest Phonofilms were limited to one ten-minute performance. Among the performers that Waddell secured for DeForest were Fannie Ward, Noble Sissle and Eubie Blake, Chic Sale, Eddie Cantor, and George Jessel. The performances of many of the artists that Waddell selected can still be seen and heard on optical or electronic projection systems today.

Five years after his work with DeForest, a notice about Waddell appeared in the *New York Times* with a brief article titled "Size of Screen." The article reported that Waddell, with John D. Elms, had been working for eight years on the development of large-format Grandeur films that would permit motion pictures to be projected on a screen forty or more feet wide. "If early motion pictures had been exhibited in theatres instead of stores there would never have been any twenty-foot-wide screens," Waddell was quoted as saying; "the natural tendency would have been to make them as wide as stages."[22]

Waddell was also described as the inventor and producer of a narrow-width endless film for home projection at the time when movies were first developed into peep shows and nickelodeon novelties. According to the article, "In Waddell's and Elms's first experiments they tried by means of two cameras to obtain stereoscopic views, but abandoned it. Then they decided that only a wider film would be logical so they came to what is now the Grandeur process."[23]

The Waddell and Elms article in the *New York Times* appeared a year after the first introduction of the large-format Grandeur process in May 1929 at the RCA Gramercy Studio in New York. Dramatic footage of Niagara Falls prompted spontaneous applause from the audience. "The great cascades seemed to splash out upon the studio floor," wrote one reviewer. "Objects appearing from the foreground to the back seemed to have a dimensional perspective."[24]

The Fox Grandeur process used 70-mm wide film with a frame four perforations high projected on a large glass screen. Waddell and Elms got some publicity for Grandeur after two feature films, *Fox Movietone Follies of 1929* and *Happy Days,* had been released in the process. *The Big Trail,* directed by Raoul Walsh and starring John Wayne, was the third feature film made with the system and was to be the last. Introduced as theaters were already going to the expense of being wired for sound, Grandeur was a visionary but short-lived and ill-timed venture.

If the reports from the historical record are true, with his innovations of small-gauge movie film for home use, the first stereoscopic projection of motion pictures in the United States, Phonofilm production, and development of the 70-mm Grandeur format, the motion picture career of William E. Waddell could best be characterized as one of consistent technological innovation.

The Lost 3-D of *Napoleon*

For his epic 1927 motion picture *Napoleon,* Abel Gance wanted to liberate the camera with a "tangible effort towards a somewhat richer and more elevated form of cinema." Addressing the audience before the premiere of his five-hour spectacle, he declared that the future of cinema was at stake. "If our language does not extend these possibilities," he avowed, "it will remain no more than a dialect among the arts."[25]

Gance attempted many experiments with cinematography in making *Napoleon,* including "Polyvision," the three-panel triptych that foreshadowed Cinerama (1952). Gance also used color and stereoscopic cinematography. Interviewed by Kevin Brownlow, Gance recalled filming *Napoleon* using both the three-camera rig assembled by Andre Debrie, his director of photography, and a second camera for color and stereoscopic photography. Camera operator Leonce-Henry Burel photographed with color and 3-D, shooting the same scenes as the triptych camera. To view the dailies, Gance used red and green anaglyphic spectacles. Gance recollected:

> The 3D effects were very good, and very pronounced. I remember one scene where soldiers were waving their pistols in the air with excitement, and the pistols seemed to come right out into the audience. I felt, however, that if the audience saw this effect they would be seduced by it, and they would be less interested in the content of the film. And I didn't want that at all.[26]

A two-color process 70-mm Fox Grandeur camera; William Waddell worked on development of the Grandeur format (1929) (photo by Ray Zone).

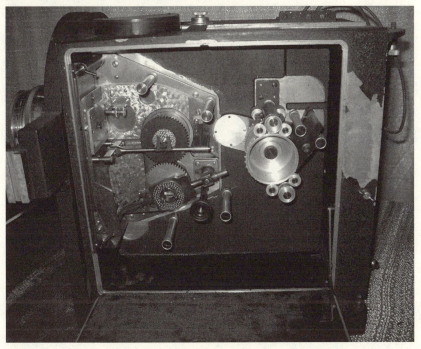

Inside the Fox two-color Grandeur camera; a two-frame pull-down is used with 70-mm film (1929) (photo by Ray Zone).

Gance shot only one roll of color film and felt that it was too late to introduce its use in the film. "Also, the 3D effect did not encourage the same feeling for rhythm in the audience," he observed. "I felt that if it fascinated the eye, it would fail to do the same for the mind and heart."[27]

Brownlow restored *Napoleon* for a full-length presentation in 1980 and gathered up every remaining film element for restoration of the epic that he could find. In a personal meeting in November 2000, I queried Brownlow as to whether he had chanced upon the stereoscopic footage during his restoration efforts. Brownlow had not. Nor had he discovered the single roll of color footage that had also been photographed by Burel at the same time.

Gance's decision to not use stereoscopic footage in *Napoleon* reflects on the medium's propensity for spectacle and display. These two qualities, as we have seen, were considered to be potentially injurious to film narrative and to the suspension of disbelief. They were also significant factors contributing to the relegation of stereoscopic cinema to prolonged novelty status.

The utopian dream of stereoscopic images in cinema, then, was a double-edged sword. The heightened realism it presented was alluring, but it had to be justified in the context of narrative.

9

Stereoscopic Cinema Proves Itself

Lumiere's 3-D Train

L'Arrivee du train was not the first film the Lumiere brothers shot in 1895 with their Cinematographe camera, but it was among the first. This powerful short film of a locomotive approaching the camera while entering a station created a sensation at its first exhibition and is one of early cinema's foundational films. It was initially photographed with a single camera and exhibited "flat" at the Grand Cafe at 14 Boulevard Les Capucines in Paris on December 28, 1895, along with other Lumiere films. It's not surprising that the Lumiere brothers later remade *L'Arrivee du train* in a stereoscopic version, but the exact date of the remake has been a subject of some confusion.

Hayes dates the stereoscopic version of *L'Arrivee du train* from 1903, stating that it was "filmed in Stereoscopic Lumiere (dual 35mm printed single strip anaglyphic)" and that it "was released in France in 1903 but never shown theatrically in the U.S."[1]

Limbacher provides two different dates on which the Lumiere brothers exhibited stereoscopic films, 1903 and 1935. "In 1903," Limbacher writes, "stereoscopic movies were shown at the Paris Exposition under the guidance of French movie pioneers, Auguste and Louis Lumiere."[2] However, in the book *Auguste and Louis Lumiere: Letters, Inventing the Cinema*, a footnote states that "Louis Lumiere had a persistent interest in 3-D images. On November 3, 1900, he took out a patent for stereoscopic moving images but it was not until 1936 that he organized public screenings of 3-D films." A chronology of the Lumiere brothers' work in this book gives 1935 as the date in which "Louis remakes 'Arrival of a Train at La Ciotat' in three dimensions."[3]

Gosser also addresses the work of the Lumiere brothers with stereoscopy. He reproduces a figure from the French Lumiere patent (no. 305,092) of 1900 with a drawing of an Octagonal Disc Stereo Device. This drawing depicts an octagonal plate mounting on a circular frame with intermittent notches, as well as a hand crank to the side. The patent states that "the images of very reduced dimensions (9mm size), are arranged in double rows on a glass plate. . . . of polygonal shape."[4] This device appears to be a form of stereoscopic kinetoscope rather than a projection apparatus. Gosser's book only covers the period 1852–1903, but he writes that "it is conceivable that they [Lumiere brothers] might have given a demonstration of their glass plate system, but so far as this author knows, the Lumieres did not achieve stereo 'film' until the 1930s, when Louis Lumiere built a horizontal run anaglyphic system."[5]

Louis Lumiere himself provided the definitive answer with an article titled "Stereoscopy on the Screen" in a 1936 issue of the *Society of Motion Picture Engineers Journal*. Lumiere illustrated with a schematic drawing and discussed an experimental device in which "the stereoscopic pair are printed upon the same film, which runs horizontally so as to produce an image practically similar to the one used in monocular vision, losing as little as possible of the sensitized surface." The device consists of an apparatus with prisms and lenses that places the left- and right-eye frames on top of each other on horizontally traveling 35-mm film. The film itself, when projected, "runs in front of two lenses, whose axes are parallel and are cut by a plane parallel to the axis, in order to allow homologous centers of the projected images to register in height."[6] The colored filters are placed in front of the projection lenses, and members of the audience are supplied with glasses of the same hues.

Lumiere conducted extensive tests to arrive at the most efficient bandwidth for the colored filters to be used in the projector and the viewers' spectacles. To obtain optimally efficient hues, Lumiere consulted the Gibson and Tyndall response curve and established the use of two filters, one a light yellow and the other a pure blue. "The difficulty in selecting the dyes," Lumiere wrote, "rested on the fact that the dyes had to show the least possible non-selective absorption and as sharp as possible limits on the edges of absorption."[7] Lumiere felt that the hues he selected would allow a viewer to watch stereoscopic motion pictures for several hours without eyestrain, experiencing striking effects, and would lead to practical application in theaters. A photo of Lumiere's stereoscopic projector in use at a February

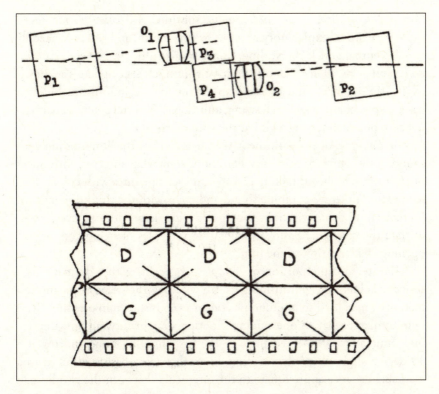

Optical system for Lumiere horizontal-run stereo movie camera of 1935 *(above);*
Lumiere arrangement for stereo pairs on the film itself *(below)* (1935).

1935 meeting of the French Academy of Sciences is reproduced in the
book *Paris in 3D.*[8]

The Lumiere brothers' final work with cinematography was three-
dimensional. Interviewed in 1948 by Georges Sadoul, Auguste Lumiere
stated, "My last work dates back to 1935, at which time I perfected a stereoscopic cinema which was shown in particular at Paris, Lyon, Marseilles
and Nice. However, my work has been in the direction of scientific research. I have never engaged in what is termed 'production.'"[9]

Audioscopiks/Metroscopiks

On January 18, 1936, *Audioscopiks,* an anaglyph novelty short with sound,
was released by Metro-Goldwyn-Mayer studios, opening for the feature

film *A Tale of Two Cities. Audioscopiks,* subtitled *A Miniature,* ran seven minutes, featured explanatory remarks by Pete Smith, and was produced by J. F. Leventhal and J. A. Norling. It continued in the same vein as *Plasti-grams* and *Stereoscopiks* with off-the-screen gimmick 3-D vignettes acknowledging the audience, and it used many of the identical motifs, including a baseball player and a girl on a swing, and closed with the perennially exciting shot of a man raising a gun at the spectators.

Audioscopiks opens with the MGM mascot, Leo the lion, roaring with sound, after which Smith's wry narration, somewhat in error, introduces the film: "You've heard talking pictures and seen pictures with color. Now, for the first time, we combine sound and color with the third dimension." Using a "nifty dame with flashy lamps," a "little lady with two eyes," to illustrate the concepts, Smith explains the principles of stereoscopy in the opening "flat" portion of the film.

The stereoscopic portion of *Audioscopiks* begins with a dramatic 3-D shot of a ladder slowly protruding from a window. The stereoscopic motion picture camera used to photograph *Audioscopiks* is shown in one shot. It's a twin camera unit with one camera shooting straight forward and the second camera at a 90-degree angle to the first shooting into a mirror at a 45-degree angle. With this configuration, the stereo base is reduced to a workable distance approximate to the interocular between the two human eyes. In another shot, of a woman holding a small puppy, stray reflections creep for a moment into the left-eye view, indicating a potential problem as a result of the mirror configuration used with strong, outdoor lighting.

Frank Nugent, film critic for the *New York Times,* wrote a somewhat humorous review titled "Meet the Audioscopiks" after a December 1935 screening:

> Mr. Leventhal was host at a four-man preview of his picture in one of the Metro projection rooms last week. He grinned happily as one spectator ducked when a pitcher wound up and tossed a baseball right out of the screen at his head. If there had been any women present, unquestionably there would have been screams when a magician conjured a white mouse onto the tip of his wand and poked it out, seemingly within arm's length of the innocent bystanders.

A brief stereoscopic segment featured a skeleton animated with stop-motion, foreshadowing Norling's work in 1939 on a film made for the New

Poster for *Audioscopiks*, which was released by MGM as a short, along with *A Tale of Two Cities* (1936).

York World's Fair. Nugent colorfully characterized Leventhal and Norling's creation as "a weird third dimension for a two-dimensional screen" and dutifully noted the necessity of holding "a strip of pasteboard with two gelatin lenses—one magenta, one blue-green before your eyes," which he characterized as "the same instrument of torture for the old stereoscopic films."[10]

Sound was used to enhance the stereoscopic illusion in *Audioscopiks*. Two segments in the film highlighted sound, with Spanish flamenco singer Eva Soba warbling a traditional song and a trombone player visually extending the sliding valve out into the audience while playing high notes. MGM first recorded their mascot, Leo the lion, roaring in 1928 and was one of the last of the major studios to build a soundstage.[11] Its first sound picture, Robert Flaherty's *White Shadows in the South Seas,* released in July 1928, had effects and music added by Douglas Shearer, a young audio engineer and brother of Norma Shearer.

By the time Pete Smith narrated *Audioscopiks,* MGM had built a soundstage in Culver City though very little sync sound was used in *Audioscopiks.* Most of the footage was shot by Leventhal and Norling in New York and at their studio on Fifty-fifth Street. Pete Smith was hired as MGM publicity director in 1925 and served in that position while the rush to sound was taking place with all the major studios.[12] For a short time in 1928, MGM hired the services of William Ray MacDonald, a speech professor from the University of Southern California, to coach their actors in diction. Also in 1928, MGM required their cameramen to attend classes taught by John Nickolaus, a Western Electric audio engineer, before allowing them to film a talking motion picture. Nickolaus subsequently went to work in the MGM camera department, working under department head John Arnold; in 1937, he built a dual-camera 3-D unit for stereoscopic filming of Smith's novelty short *Third Dimension Murder.*

In a 1939 article in the *Society of Motion Picture Engineers Journal,* J. A. Norling wrote that "*Audioscopiks* was one of the most successful shorts ever made" and that its "success is a sure indication of public acceptance" for wearing red and green spectacles in viewing a film.[13] *Audioscopiks* and its two successors, *New Audioscopiks* (1938) and *Third Dimension Murder* (1941), were among the most widely seen stereoscopic motion pictures of all time. To prepare for the release of *Audioscopiks,* MGM manufactured 3 million red and green lorgnette spectacles at a cost of $3.25 per thousand.[14]

For release prints of *Audioscopiks,* MGM hired the Technicolor Motion Picture Corporation. By 1935, Technicolor had created the three-strip color process used in the feature film *Becky Sharp* released that same year. *Audioscopiks,* however, was released in anaglyph and only required a two-color process. Technicolor had introduced their two-color process using dye imbibition in 1919. "A first approximation of the Technicolor imbibition method consisted of two gelatin reliefs produced upon thin celluloid which were glued or welded together back to back and dyed in complementary colors," wrote H. T. Kalmus, director of Technicolor. "Combined with the Technicolor two-component cameras, this method provided an immediately available system (1919–21) capable of yielding two-component subtractive prints."[15] Throughout the 1920s, for the anaglyphic novelty shorts, *Plastigrams,* and the *Stereoscopiks,* release prints were made by William Van Doren Kelley's Prizmacolor Company, which used gelatin reliefs printed on both sides of a single strip of film.

The early two-color Technicolor feature films *The Toll of the Sea* (1922) and *The Black Pirate* (1925), released with the double-coat gelatin reliefs on two cemented positives, created a lot of difficulties for exhibitors. The need for a two-color imbibition process with the dyes on only one side of a single strip of film was obvious. "Our troubles never ended," wrote Kalmus; "we brought out two-color imbibition prints with silver sound track in 1928. The advantages in respect to focus, cupping, scratching, size of reel, and cost of manufacture were immediate."[16] This was the release format for the MGM anaglyphic shorts.

Pete Smith worked as MGM publicity director until 1930. Then Smith began to write, produce, and narrate a series of one-reel shorts known as the "Pete Smith Specialties," the first of which, *Fisherman's Paradise,* appeared in 1931. Smith's Specialties humorously touched on every conceivable subject from sports, technology, and parenting to marriage, motion pictures, and shopping. From 1931 to 1955, Smith produced over 150 short films, sixteen of which were nominated for Academy Awards with two of them winning Oscars.

In a 1953 article titled "Three Dimensionally Speaking," Smith wrote (somewhat hyperbolically) about his experience producing the three MGM stereoscopic short films:

> In the early part of 1936 MGM introduced something so new, so startling, that audiences throughout the country could hardly retain their seats when they saw it. Upon entering the theatre they had been given a pair of celluloid glasses with red and green lenses,

with the admonition not to throw them away. After draping themselves in the plush seats the patrons were suddenly stirred from their placidity by objects that seemed to jump out of the screen and right into their laps as the tenuous tenor of your scribe boomed from the far reaches of the sound track. Here was a brand new thrill, a complete change in the presentation of motion pictures.[17]

Smith recounted that Fred Quimby, head of the MGM Short Subjects Department, had purchased stereoscopic footage from Leventhal and Norling, then "said footage was turned over to me for editing and sound treatment and narration." In this way, both *Audioscopiks* and *New Audioscopiks* were finished by Smith. The two films were extremely successful. "Pardon my immodesty," noted Smith, "but the Sales Department assured me that these pictures played to maximum bookings throughout the world. As a result of such success, we at MGM decided to enter the 3-D field."[18]

Starting from scratch, Smith enlisted John Nickolaus to build a twin-camera stereoscopic rig at a cost of about $3,000. Nickolaus used two conventional Bell and Howell cameras with specially matched lenses, 2¾ inches apart, on a baseplate in a lens-to-lens configuration. Assisting Nickolaus were technicians Irving Ries and A. G. Wise. Smith also worked with director George Sidney—who was later to direct *Kiss Me Kate* in 1953 in 3-D for MGM—in a basement on the studio back lot.

The new MGM 3-D rig, dubbed "Metroscopix" by publicist Howard Dietz, was used to photograph "a travesty on the Frankenstein craze of that era. Unlike the first two subjects, which were a series of clips, our picture told a complete story."[19] The story of *Third Dimension Murder* related the narrator's encounter with the Frankenstein monster at an old castle, along with a witch and a wooden Indian who came to life. MGM's Frankenstein monster, portrayed by Ed Payson, was similar to Universal's classic version but slightly reworked. Photographed with excessive parallax by Walter Lundin, the Frankenstein monster ran amok in the castle and, as a finale, hurled a variety of objects from atop a wall down at the Metroscopix camera standing in for the audience.

With his article on three-dimensional motion pictures, Norling wrote with clarity about stereoscopy and made a strong defense of the anaglyph process:

It has been claimed that the anaglyph system, because one eye sees one color and the other its complementary, causes serious retinal rivalry in a short time. Personally I have never noticed anything of the sort, and I have looked at anaglyphs for hours on end. Any defense of the system based on purely personal experience may be waived, but it has been my observation that no one suffers any discomfort in viewing a reel or two.

Norling was also definitive in championing anaglyph as an effective form of stereographic display. To the criticism that anaglyphs presented a sensation of depth as a "series of thin cardboard models placed in different planes" and that they "lack roundness," he responded that "the author feels that that is ridiculous and that no one who has viewed good anaglyphs can honestly make such a criticism."[20]

Despite these laudatory comments, Norling conceded that polarizing image selection is a superior form of stereographic display. It made projection of full-color images possible and afforded a "more complete elimination of the unwanted image."[21] With the inauguration of polarizing image selection in the 1930s, the stereoscopic motion picture would come of age. This classic format—dual-projection with polarizers crossed at right angles to each other—would serve as the platform on which the golden age of 3-D motion pictures in the 1950s would take place.

Polarizing Image Selection

A ray of light has a direction of vibration. A light polarizer determines and selects the direction of vibration of a ray of light. As early as 1852, William Bird Herapath, an English physician, discovered a synthetic crystalline material that transmitted full-color polarized light with high intensity. Chemically, Herapath's material was made of minute crystals that were composed of sulfate of iodo-quinine. Sir David Brewster was very interested in Herapath's work, hoping that it could be used in microscopes and kaleidoscopes. Herapath's crystalline structures, however, were so unstable they would shatter quite easily, and commercial development proved impossible. These fragile crystalline plates eventually became known as herapathite.

Edwin H. Land was fascinated by Herapath's work with crystals. "Herapath was deeply impressed to observe that in some places where the crystals overlapped they were white," Land explained, "and in other places where they overlapped they appeared a deep blue-black. He realized that these were polarized crystals."[22] Building on Herapath's work, Land filed his first patent application in 1929 for a sheet polarizer, making polarized image selection for stereo projection a practical possibility. Land used flexible, plastic film with needle-shaped crystals of submicroscopic size that aggregated and oriented uniformly on the plastic sheet. The patent (U.S. no. 1,918,848) granted on July 18, 1933, described an invention in which "a refracting polarizing body is produced by forming a colloidal suspension containing a mass of relatively small polarizing bodies which are oriented in a field of force so that the polarizing axes of the polarizing bodies are in alignment, or substantially parallel."[23] Precision, close tolerances, ease of use, and manufacture for large-scale production were made possible with Land's process.

From the beginning, Land viewed stereoscopic motion pictures as a potential application for the polarizing material. Another important application was the reduction in automobile headlamps and windshields to reduce headlight glare and improve highway safety.

Land's polarizing material was first used for projection of still stereoscopic images at the behest of Clarence Kennedy, an art history instructor at Smith College who wanted to project photo images of sculpture in stereo to his students. On March 22, 1934, Kennedy wrote that "the two young inventors, who have their laboratory in Boston, gave me, on my own machines here at the College, a convincing demonstration of the practicality of studying from stereo-projections on a screen."[24] Kennedy subsequently published an article in the *Society of Motion Picture Engineers Journal* on the educational benefits of stereophotography and projection.[25]

George Wheelwright III, who was very interested in 3-D movies, had entered into a business partnership with Land. Together, they would name the new polarizing film Polaroid and jointly form the Polaroid Corporation. With Wheelwright, Land began to experiment with stereoscopic motion pictures. The pair used several methods of stereoscopic production, including dual-camera dual-projector methods, beam-splitter, and alternate frame techniques.

In 1934 Wheelwright and Land visited Kenneth Mees, photo research director at Kodak, bringing along fifteen minutes of stereoscopic movies with them. Subsequently, Wheelwright and Land were supplied with Kodak's new Kodacolor film. With a dual 16-mm format, Land and Wheelwright produced the first stereoscopic motion pictures in color in 1935. William H. Ryan wrote:

A later stereo motion picture made in a standard camera equipped with a beamsplitter of Land's design and projected in a standard

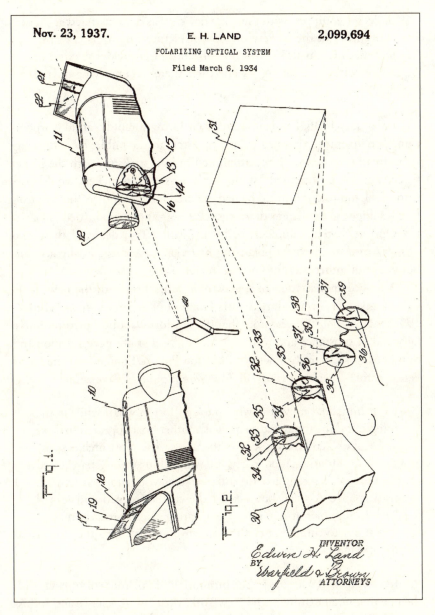

Nov. 23, 1937.

E. H. LAND

POLARIZING OPTICAL SYSTEM

Filed March 6, 1934

2,099,694

The two uses for Polaroid film proposed by Edwin H. Land in "Polarizing Refracting Bodies": fig. 1, reduction of headlamp glare; fig. 2, stereoscopic movies (1933).

projector equipped with the same device, was demonstrated at the convention of the Society of Motion Picture Engineers at Washington, D.C., in 1935. This novel beamsplitter divides the field of a lens without dividing its aperture, thereby conserving all the light.[26]

In January 1936, Land announced the formation of Polaroid Corporation. The inventor explained polarizing filters to the public by comparing them to a picket fence. If one attempted to hurl sticks through the picket fence, only those oriented on the same axis as the pickets would pass through. Land told the world he had material for 3-D movies and controlling headlight glare. Patent drawings for a new application titled "Polarizing Optical System," applied for by Land in 1934, depicted these two primary uses for the new polarizing material.[27] Land eventually also announced 800 other uses for Polaroid material.

After a demonstration of stereoscopic projection with the new material was given at the Waldorf-Astoria Hotel in New York City in February 1936, the *New York Times* wrote that it "transforms motion pictures into a fairyland of substance and reality."[28] After Wheelwright gave a demonstration in May 1936 to the Society of Motion Picture Engineers at the Hotel Pennsylvania, a reporter from the *New York Herald Tribune* wrote:

> [A] fishing boat moved slowly up to [a] Boston wharf while in the foreground the ropes and rigging of another boat appeared so close and real one could almost seize them. . . . The most lifelike scene was a children's garden party, in which four youngsters seated about a table scrambled for a basket of favors wrapped in pink tissue paper. A small boy nearest the camera turned and leaned back in his chair at one juncture, creating the impression that he was about to tumble out of the screen "window" into the laps of the spectators.[29]

In December 1936, the public promotion of Polaroid culminated with the opening of an exhibit at the New York Museum of Science and Industry at Rockefeller Center. *Literary Digest* magazine reported that visitors to the exhibit could "see for the first time a new scientific weapon which cracks down on errant light waves, promises to slay the dragon of headlight glare." After entering a room where two powerful headlights beamed down on them, visitors stepped behind a glass windshield that dramatically reduced the glare to two purple disks. Then, exhibit patrons moved into another room. "Museum visitors also don polarizing spectacles, gaze upon polarized movies which leave them gasping with astonishment," wrote *Literary Digest:* "Before their eyes, the movie-screen vanishes, becomes a window through which they look at action in startling third dimension. So real is the illusion of depth that occasional spectators reach out to touch objects on the screen."[30]

The Polaroid exhibit remained at the Museum of Science and Industry for several years, and the 3-D film was shown several times daily to many thousands of visitors. Land's proposed use of Polaroid material in headlights and windshields never came to fruition. The use of Polaroid material for stereoscopic motion pictures, however, was just beginning.

Zeiss Ikon Stereo

The Zeiss Ikon consortium of camera companies in Dresden, Germany, was formed in 1926, and stereoscopic photography played a significant role in the product lines. The first Zeiss Ikon catalogue of 1927 included twenty different stereoscopic still cameras, along with variants for each model, including single-lens reflex (SLR) stereo cameras, as well as stereo drop bed and strut cameras.

Zeiss Ikon began producing stereoscopic motion picture cameras and projectors in the 1930s for both dual-strip and single-strip 3-D films. Working in cooperation with the State Establishment for Physics and Technology of Braunschweig, Zeiss Ikon developed a high-speed twin 16-mm stereo motion picture film system for use at the Berlin Olympics of 1936. The two 16-mm cameras were mechanically linked to stay synchronized.

In 1935, Otto Vierling developed a single-strip 35-mm stereo camera system using a prism in front of the lens for Zeiss Ikon. Frank Weber observed that Vierling's 3-D shooting method was based on what is called "the 70 arc-minute condition." It was demonstrated at the German Society for Photographic Research in June 1936 with black-and-white, silent film. Weber wrote that the polarizing filters installed on the projector and in the spectacles the audience used were manufactured by the Zeiss Company on a commercial basis in 1936; "independently, at about the same time, they were also made in Germany by E. Kasemann and by J. Mahler."[31] Mahler later moved to the United States, and, working with Edwin Land, proved instrumental in developing the Vectograph process.

Zeiss polarizing glasses were also used for viewing *Beggar's Wedding* (*Nozze Vagabonde*), the first stereoscopic feature with sound, shot in black and white in Italy in 1936 with the Gualtierroti stereo camera for the Italian Society of Stereocinematography. Eddie Sammons described the plot:

> About to end his bachelor life, an actor is given a supper by his colleagues to see him off into marriage. Events take a strange course at the party when a real £1000 note, indistinguishable from the one used by a conjurer is burned. (Generally noted as being presented in a polarized system, it [*Beggar's Wedding*] stands in film history as the first feature-length talkie in 3-D . . .).[32]

A reviewer for *Cinema* magazine wrote that *Beggar's Wedding* was "a waste of ingenuity because a variety of ideas flare up only to go out without succeeding in adding to the film drama."[33]

Subsequently, Vierling developed three different configurations of the Raumfilm-Verfahren single-strip stereo system for 35-mm film cameras with a beamsplitter prism. Two side-by-side vertical images were captured on a film frame in the first format. The second used horizontal images, 11 by 14 mm, turned 90 degrees, and the third used two head to foot images turned 90 degrees.[34] Light rays going into the camera passed five reflecting surfaces in front of the prime lens, and the normal focal length for the lens was usually 40 mm.

The first color stereoscopic motion picture with sound, *Close Enough to Touch* (*Zum Greifen nah*), a short commercial film promoting insurance, was shot with Vierling's system on Agfacolor film. It premiered at the Berlin Ufa Palast theater on May 17, 1937, and was shown with a custom attachment which "de-rotated" the images on a standard 35-mm projector. Zeiss Ikon polarizing filters with the trade name Herotar were used for stereo projection. *Close Enough to Touch* was photographed in Dresden and has not been shown in the United States.

In 1939, Zeiss Ikon produced a twelve-minute self-promotional single-strip stereoscopic 35-mm sound film with Agfacolor titled *Six Girls Roll into the Weekend* (*Sechs Madels rollen ins Wochenend*). It was never released in the United States.

The German military used the 35-mm Raumfilm-Verfahren stereoscopic film system to photograph about 300,000 feet of film footage during World War II. Some of this footage, consisting primarily of training

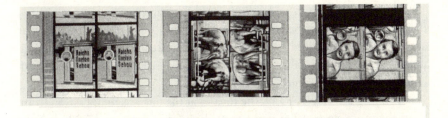

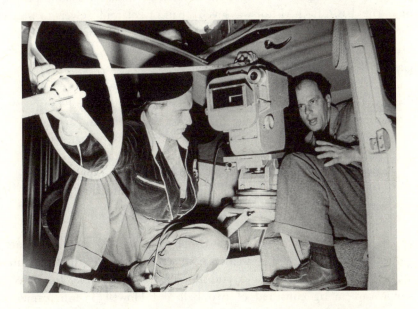

Three versions of the 35-mm single-strip Zeiss Ikon stereo system *(above);* Zeiss Ikon stereo camera used for Volkswagen film *(below)* (1953) (both from Walter Selle Archives).

films, was recovered in 1994 from the ruins of a church that was bombed in Dresden and copied from nitrate stock to modern film material by the Cinema Film Division of the Federal Archive at Berlin.

Stereo Films at the Fairs

In 1939, the Loucks and Norling studios produced *In Tune with Tomorrow,* a 35-mm black-and-white stereoscopic motion picture for the Chrysler Corporation exhibit at the New York World's Fair. This twelve-minute film was, as John Norling characterized it, "the first full-size Polaroid three-dimensional picture ever to be made" and "the first public showing of really satisfying three-dimensional pictures."[35] Norling's efforts on *In Tune with Tomorrow* produced a real advance for stereoscopic cinema. Writing about the World's Fair films, Richard Griffith observed that it exemplified a "valid use of film trickery."[36]

Opening on May 4, 1939, the theater at the fair was in operation from 10 in the morning to 10 at night and admitted as many as 17,000 people on a daily basis. Each visitor was given a pair of Polaroid viewers to watch the twin-strip 3-D movie shown, which used two projectors mechanically interlocked by a selsyn motor drive on a silver screen. Through the 3-D glasses, Norling wrote, the audience saw

> actual operations in the half-mile long Plymouth plant at Detroit, and a trick-assembly picture that shows the Plymouth car magically coming into being without the apparent aid of human hands. The various parts of the car, numbering many thousand, come waltzing in together or separately, each to take its proper place in engine, chassis or body—all in carefully synchronized step to the beat of the music.[37]

It took thirty-six days to photograph over 13,000 motion picture frames using stop-motion animation with separate parts of the Plymouth moved a predetermined distance and shot one frame at a time. The stereoscopic camera rig consisted of two Bell and Howell 35-mm cameras mounted so that one was upside down. With this configuration, the two lenses could be brought sufficiently close together for an interaxial separation of 3¼ inches. Norling calculated that an interaxial of 1½ inch would have been preferred, but he compensated for this deficiency by using matched lenses of different focal lengths for a satisfactory compromise.

For filming of the auto assembly sequences, three different stages were erected at the Plymouth plant in Detroit. Parts that moved along the floor were relatively easy to film but required precise timing and spacing. Making an entire motor or auto body appear to sail through the air was a technical challenge. The large auto parts were suspended by steel wires from a movable trolley that ran along an overhead track. The wires were painted to make them invisible and were plucked like guitar strings at the moment of filming so that the vibration camouflaged their presence.

The narrative order of scenes in the film followed the sequences of automotive assembly as it was actually done. Editing and timing was done during shooting, with each foot of film predetermined for music and sound effects cues. A special listening machine was used during filming with beats and musical notes marked for cues, each one of which was given a frame number. Charts and exposure sheets depicted the movement of objects through space, with frame numbers for each stopping point. The overhead track was marked in inches and feet, and the trolley carriage was fitted with a ratchet device to raise or lower auto parts suspended by wires.

About half of the movie was photographed backward, with parts removed out of frame. With film running reversed, the objects then appeared to move into the scene. The twin-strip stereoscopic projection was on a screen 17 feet wide that had been sprayed with aluminum coating. The aluminum lacquer contained no pigment paint.

In 1940, a Technicolor version of Norling's Chrysler film, titled *New Dimensions,* replaced the black-and-white version at the New York World's Fair. For the stop-motion in *New Dimensions,* the interlocked Bell and Howell cameras were each fitted with a three-color filter wheel for making the stereoscopic color-separation negatives with a sequential exposure method. Black-and-white 35-mm panchromatic negative film was used. Norling wrote:

> Color balance was attained by making sectors having angular dimensions calculated to pass the quantity of light required for each color and as demanded by the sensitivity of the film. The "A" (red) filter passed light to which the film was more sensitive than that passed by the "B" (green) and "C5" (blue) filters. Consequently, the red filter had the narrowest opening of all and the "C5," to whose transmission the film was least sensitive, had the widest opening.[38]

The successive frames were then step printed to three strips of negative film. Technicolor then used the red, green, and blue separation negatives to make printing matrices for dye imbibition prints combining all three colors on a single strip of film for projection.

It was at this time that Norling constructed his dual-band twin-lensed stereo camera housed in a single body. The rackover camera had a variable interaxial separation system ranging from 1½ to 4 inches. A binocular viewfinder was used to make interocular adjustments and set convergence,

Souvenir map and Polaroid 3-D glasses for the New York Worlds' Fair (1940).

making a three-dimensional visual inspection of the shot possible before rolling films.

Loucks and Norling also produced another twin-strip 35-mm black-and-white three-dimensional film with *Thrills for You,* a major attraction at the Golden Gate International Exposition in San Francisco in 1940. *Thrills for You* was photographed by J. F. Leventhal and produced for the Pennsylvania Railroad exhibit at the exposition. This promotional film featured exciting stereo cinematography, depicting onrushing locomotives and footage shot from the front of a rapidly moving train, as well as interior views inside different rail cars.[39]

With an address to the Society of Motion Picture Engineers in May 1941, Norling estimated that about 4 million people had seen the Chrysler and Pennsylvania Railroad 3-D films and conjectured that "it is probably safe to say that real three-dimensional motion pictures have emerged from the experimental and novelty stage."[40] In reality, the novelty period for stereoscopic cinema would continue for another decade, but not without significant developments.

Vectographs

Collaborating with Clarence Kennedy and Joseph Mahler in 1940, Edwin Land developed the Vectograph, what his publicist, Richard Kriebel, characterized as "a new kind of pictorial representation. At its most spectacular, it is a three-dimensional picture in the form of a single print."[41] John Norling described the Vectograph very simply: "In this new film the image itself is a polarizing image. It polarizes the light passing through and does so in varying degrees."[42]

The Vectograph combined the left- and right-eye images, along with polarization crossed at 90 degrees, on a single print. Initially developed for still stereo images, the Vectograph was initiated by Czech inventor Joseph Mahler, who immigrated to the United States when Germany invaded Czechoslovakia in 1938 and was hired by Polaroid. Land and Mahler received a patent in June 1940 for an "Apparatus Employing Polarizing Light for the Production of Stereoscopic Images" as a "superimposed plurality" of images.[43] Land described the critical element in the process, explaining that "three dimensional pictures could be obtained if the two images of the stereoscopic pair were made in terms of percent polarization."[44] The darkest areas of an image result from almost complete polarization, lighter areas are partially polarized, and white areas carry almost no polarization.

In an article in the June 1940 *Journal of the Optical Society of America,* Land officially named the Vectograph:

> The purpose of this paper and others to be published later is to discuss the valuable properties of a *controlledly non-uniform* vectorial field created with or in polarizing surfaces. This new kind of field when it forms an image will be termed a vectograph; thus a vectograph is an image rendered in terms of vectorial inequality.[45]

In 1935, while still a resident of Czechoslovakia, Mahler was granted a U.S. patent for a "Stereo Apparatus" consisting of "two optical projection systems placed at right angles to one another, one for each of the two stereo-diapositives."[46] In 1946, Mahler applied for a patent incorporating Vectograph film; this was a simple and portable backlit "polarizing stereoscopic transparency viewing apparatus."[47] This patent was eventually granted in 1951.

The Vectograph was used during World War II for aerial reconnaissance, when Land worked with the U.S. military. Although the war delayed commercial exploitation of the Vectograph for the consumer market, after the war, considerable work took place at Polaroid to develop motion picture Vectography. Experimental Vectograph motion pictures were made in both black and white and color.[48] The Vectograph held great promise to simplify stereoscopic motion picture projection. A single projector running Vectograph motion picture film, with a silver screen and the audience wearing Polaroid glasses, would have eliminated many of the difficulties and errors that can arise with both twin-strip and single-strip stereoscopic projection.

The Polaroid Corporation remained committed to three-dimensional pictures and the Vectograph. When the David White Company introduced the Stereo-Realist Camera in 1947 to the amateur photography community, Polaroid Corporation supported the market by making available circulars with lists of resources and suppliers of stereoscopic materials and equipment. A five-page June 1951 Hectograph circular from the Polaroid Corporation, for example, was titled "Information about Three-Dimensional Pictures" and included supplier addresses for three-dimensional projectors, cameras, and silver, nondepolarizing, screens.[49]

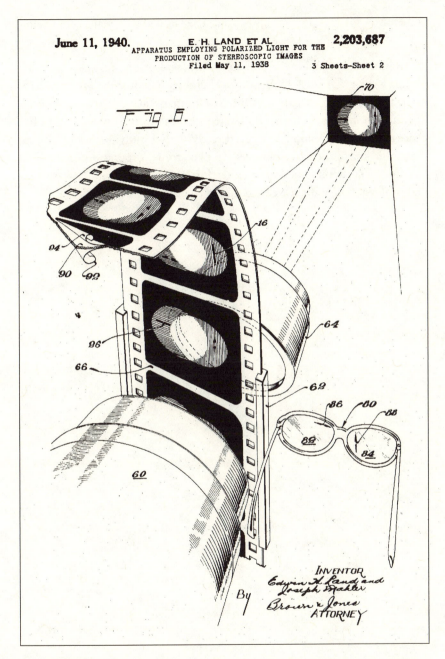

June 11, 1940.   E. H. LAND ET AL   2,203,687
APPARATUS EMPLOYING POLARIZED LIGHT FOR THE
PRODUCTION OF STEREOSCOPIC IMAGES
Filed May 11, 1938   3 Sheets—Sheet 2

Dramatic patent illustration of stereoscopic projection with Edwin Land and Joseph Mahler's Vectograph, "Apparatus Employing Polarizing Light for the Production of Stereoscopic Images" (1940).

Resources for making Vectograph slides and prints were also included in the circular. A 2 by 2–inch custom Vectograph glass slide, priced at $15, was available from the Stereo Optical Company at 3539 North Kenton Avenue in Chicago, Illinois. The Stereo Optical Company also sold film, processing materials and chemicals directly so that individuals and companies could make their Vectographs. The Polaroid Corporation understood "seeding" and supporting the market for stereo photographers, amateur and professional, and the Vectograph.

10

The Stereoscopic Overture Finishes

Autostereoscopic Cinema

Autostereoscopic cinema has always been a "holy grail" for utopian inventors, an extremely exotic and challenging subset of stereoscopic motion pictures. Many have foundered in its quest. In a November 1931 issue of the *New York Times,* an article appeared titled "New Screen Gives Depth to Movies" about cinematographic engineer Douglas F. Winnek and his stereoscopic projection of motion pictures, "which, it is said, makes possible a perspective of three dimensions without the necessity of any special viewing devices."[1]

Winnek's stated method used a new type of screen with a beaded cellophane surface, with each bead acting as a lens. Some 576 beads covered 1 square inch, and a standard binocular projection of left- and right-eye images purportedly separated the two views for binocular steropsis. In the same article, mention was made of Dr. R. T. A. Innes, an astronomer from South Africa, who "had perfected a system of stereoscopic film projection founded on a special optical appliance fitted near the screen."[2]

The autostereoscopic efforts derived from F. E. Ives's foundational patent of 1903, "Parallax Stereogram and Process of Making Same," which made stereoscopic images viewable without any optical device.[3] It worked by alternating the left- and right-eye views vertically on a transparent line screen, using opaque lines with clear spaces between them. The line screen consisted of vertically running cylindrical lens elements, lenticules, so that these screens were typically called lenticular. With an 1899 patent titled "Pictorial Reproduction," John Jacobson had propounded the fundamental principle of alternating clear and opaque lines, but his patent was insufficiently detailed and most likely never reduced to practice.[4]

Frederic Ives's son, Herbert, of the Bell Telephone Laboratory, was an assiduous worker in the field of autostereoscopic cinema. On October 31, 1930, he presented his apparatus for autostereoscopic motion pictures in a first public demonstration to the Optical Society of America. According to an account of it in the *New York Times:*

> The pictures shown were small and on account of the experimental nature of the machine used, were visible only to small groups at a time. Dr. Ives said that the methods were susceptible of refinement and theoretically applicable to the production of motion pictures in relief, although at a cost now which would be practically prohibitive for commercial purposes.[5]

In the April 1932 issue of the *Society of Motion Picture Engineers Journal,* Herbert Ives wrote about the problems of projecting motion pictures in relief. By this time, he had begun to experiment with using a battery of projectors, as many as thirteen, for autostereoscopic projection upon a lenticulated screen. The disadvantages of this method, Ives wrote, "are the excessive refinement of all the apparatus parts, which could be avoided only in part by having recourse to a multiplicity of projecting units or excessive speeds of projection."[6]

On April 28, 1933, Ives demonstrated a further refinement of autostereoscopic movies with added complexity to 500 members of the Society of Motion Picture Engineers in New York City. He had solved certain problems, but his ultimate prognosis was not hopeful, as he declared that there were "conditions which, in his opinion would keep motion pictures in depth as a laboratory curiosity forever, without practical application in a motion picture theatre." Close-ups were an impossibility, and Ives's means of production were incredibly complicated. He finally settled on a "device of photographing an object as seen in a curved mirror, to provide diversity of viewpoints, and recording it on a sensitized plate with convex ridges running from top to bottom on the top and back." Projection was equally complex: "The film must not shift by as much as one-hundredth of the width of a picture strip, or 1–50,000 of an inch in passing through the projector. He obtained this only by printing his picture strips on a disk which limited his total footage to the circumference of a disk about four feet in diameter."[7]

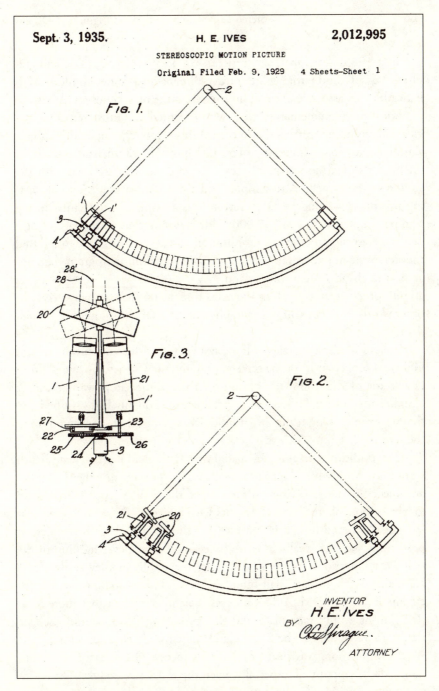

Herbert Ives's exceedingly complex autostereoscopic system, titled "Stereoscopic Motion Picture" in the patent (1935).

In a 1935 *New York Times* article, "Camera in Revolt," cinematographer Gregg Toland wrote about a three-dimensional camera built by William Alder, an associate of the California Institute of Technology. Alder was a motion picture cinematographer who invented an autostereoscopic device with spinning mirrors that recorded three different angles of a subject simultaneously on a single strip of 35-mm film. Alder's small attachment fit on the lens of a motion picture camera and was synchronized with the shutter of the camera. A tiny motor and a prismatic arrangement of turning mirrors, revolving at 2,000 revolutions a minute, ran silently during filming. Built at a cost of $52,000, Alder's system flew in the face of binocular stereopsis, and the inventor himself said as much. "I have found that the stereopticon photographs that are made with still photography are not at all true vision," Alder told Toland. "By the use of two pictures, one gets an illusion of greater depth in the third dimension than actually exists. In other words, stereopticon photography creates a false illusion of too much depth."[8]

Toland shot test footage with the system and with movie producer Sam Goldwyn looked at it on the screen at Grauman's Chinese Theater. "From any sector of the house the illusion of three dimensions was startlingly clear," wrote Toland; "it seemed as if the screen suddenly had vanished and that we were looking really into a solid, live mass of flesh-and-blood upon the screen."[9]

The traditional techniques of lighting with shadows to create depth would not work with Alder's process. It was important to light all sides of an object for the revolving mirrors to see those sides. And it would no longer be possible to flatter the actors with lighting. Toland observed that "this new demand in lighting means that those stars whose bad noses, or protruding ears, or weak chins we have favored by clever lighting in the past must now stand up and be seen by the public as they really are."[10] Painted backgrounds and half-built houses would also no longer suffice for cinematic illusion with Alder's process. Toland predicted that more location filming would take place and that production costs would rise as a result. There would also be less necessity for close-up shots of actors.

Nothing more was heard of Alder's process. His investors were undoubtedly disappointed, especially given the lack of a product after an investment of $50,000. Any system dispensing with binocular stereopsis, however, could not work. That is the single incontrovertible fact at the heart of three-dimensional images. It is surprising that Toland, one of the great lighting cameramen, thought that he was actually seeing a stereo-scopic image at the Chinese Theater.

Five years after experimenting with Alder's system, Toland used light and shadow with deep-focus cinematography to create great monocular depth onscreen in the filming of Orson Welles's *Citizen Kane*. In a highly detailed article published in a 1972 issue of *Screen* magazine, Patrick Ogle wrote that "Toland's name has become synonymous with deep focus cine-matography." Ogle surveyed the evolution of film stocks, taking into ac-count varying speeds and the practice of using smaller apertures on the camera to increase depth of field and sharp focus throughout the image. Given cinema's foundation as a technological art, Ogle found it surprising that most scholarly work about motion pictures "seems almost consciously to eschew any concern for film technology and the relationship it bears to many aesthetic and historic trends in the cinema" and suggested that ste-reoscopic film may be an extension of deep-focus cinematography.[11] Gosser argued to the contrary, writing that "the development of 'deep-focus' cin-ematography may have also helped to supplant 3-D film."[12]

Throughout the 1940s, there were continuing attempts to create au-tostereoscopic cinema displays. Suzanne Carre in Paris invented a system with a reciprocating grid made of wires or rods placed in front of a conven-tional motion picture screen and moving rapidly back and forth. The stereo pair of images was alternately rear-projected on the moving grid, all of which was running in sync with the shutter of the projector. In 1948, Francois Savoye originated the Cyclostereoscope, which uses a rotating grid for stereoscopic selection in the form of a truncated cone that revolves around the projection screen. The Cyclostereoscope was successfully shown in 1949 to an audience at Luna Park in Paris. The limitation with Savoye's process, as with all autostereoscopic displays, was the limited viewing zones for seeing 3-D, usually constricted to an angle of about 40 degrees.[13]

Semyon Ivanov—Utopia at Stereokino

Russian filmmaker Semyon Ivanov expanded on the principles that F. E. Ives had established with the parallax panoramagram, and in the 1940s in Russia he actually made autostereoscopic motion pictures that worked. In 1941, Ivanov produced a short parallax stereogram motion picture titled *Land of Youth* and subsequently wrote about it for *American Cinematogra-pher* magazine. Regarding filming, he described the use of an ordinary mo-

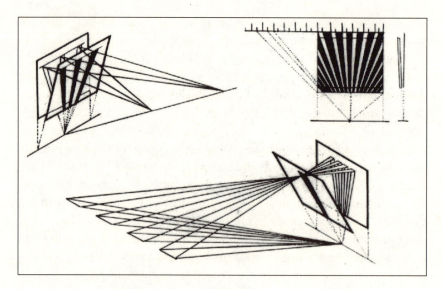

The Ivanov parallax stereogram system (1948).

tion picture camera equipped with a simple device "called a stereo-nozzle, consisting of two mirrors connected by a hinge and placed at an angle somewhere approaching 180 degrees to each other, [which] is placed in front of the objective of the cine-camera. These two mirrors divide, as it were, the one image into two that are fixed on the film."[14]

In December 1940, the Moscow Stereokino, a special theater with 180 seats, was constructed for viewing these films. The screen grid consisted of a large number of regularly spaced parallel copper wires stretched from top to bottom. The stereo films were back-projected over an area about 14 feet high and 19 feet wide. Ivanov described the unique screen:

> The stereoscopic screen is formed of a metal framework weighing six tons. Over this thirty thousand copper wires of a total length of a hundred and fifty kilometers (about 93 miles) are drawn, forming a "perspective grille."
>
> The wires are so fine and so close that they cannot be distinguished one from the other at a distance of ten metres (about 33 ft.) They are, of course, fitted with the greatest mathematical precision, to the hundredth part of a millimeter.[15]

In 1941, Ivanov completed filming of *Robinson Crusoe,* the world's first feature-length autostereoscopic motion picture. *Robinson Crusoe* was photographed on 70-mm film with side-by-side stereo images having an aspect ratio of 1.37 to 1.

Autostereoscopic movies played at the Moscow Stereokino for eighteen years, and four additional theaters were built in Russia to use the grid screen. Two more feature films were reputedly produced in the parallax stereogram process: *Lalim* (based on a Chekhov story) and *Aleko* (from a poem by Pushkin). Ivanov kept tinkering with the projection screen and eventually experimented with both an embossed lenticular sheet and a glass screen using etched and ruled vertical lines.

Ivor Montagu, in his book *Film World,* has described the limitation of Ivanov's parallax stereogram movies: "The disadvantage of this system is that you must sit just so, for your two eyes to see the two images as one. A slight shift of the head and their coincidence is lost and you must wriggle in your seat until you find them 'right' again."[16] Even so, many reviewers of the films were enthusiastic. About *Robinson Crusoe,* one of them wrote: "Out in the auditorium, about three rows in front of you, leaves and lianas materialize in the air, dangle and dance, and float away into Crusoe's face."[17]

After seeing Ivanov's work, the great Russian film director Sergei Eisenstein wrote: "To doubt that stereoscopic cinema has its tomorrow is as naive as doubting whether there will be tomorrows at all."[18]

Single-Strip 3-D Systems

Otto Vierling was not the only developer of single-strip 3-D motion picture systems. Throughout the 1940s, several different inventors attempted to simplify stereoscopic photography and projection by capturing left- and right-eye images on a single strip of 35-mm film.

Chief among the promoters of single-strip 3-D systems was Frank A. Weber, in Holland, whose system was dubbed "VeriVision." According to Weber:

> Here in Europe, where stereofilm development has a long history, it is now generally accepted that both stereotaking (the photography) and stereoprojection should be done by means of a single film. This insures true position of the left and right images in relation to each other, easy printing on existing types of film printers, and a commercially acceptable method of projection.

Because of the 60 percent loss of light that occurs with polarizing filters, Weber suggested that the images in a single-strip system should be full-aperture dimensions: 16.03 mm by 22.05 mm for 35-mm film, and 7.42 mm by 10 mm for 16-mm film. Weber insisted on very small "stereobases (interocular distances) and stereoangles (the angle between the two optical axes going from camera to object)" to avoid excessive distortion.[19]

In 1948, using these formulas, Weber claimed to have "produced in VeriVision what he believes was the first 3-D newsreel in color. It showed a procession passing before Queen Juliana following her ascension to the throne."[20] For some reason, Weber then printed the footage on two 16-mm filmstrips for projection. He recorded that the footage was first exhibited in 1949 in the Hague in Holland and in 1951 in London. "To be universally acceptable," Weber wrote, "a stereofilm camera should offer stereobases from as small as 1/4-inch up to about 8 inches or more, and stereoangles from 0.3 inch to 3 inches or more."[21]

Weber applied for patents in various countries for his universal stereo-film camera, which he claimed could be made by modifying any standard 35-mm motion picture camera. He described the four essential changes necessary: (1) modification of the film transport mechanism for a two-frame pull down in one stroke of 38 mm; (2) doubling in size of the film gate aperture; (3) turning of the camera viewfinder 90 degrees to match the camera, which is turned sideways 90 degrees for filming; and (4) replacement of single lens mount with new twin lens mount, one lens above the other.[22] For projection of VeriVision, modification to standard projectors was necessary. These modifications included use of sprockets of double diameter to produce twice-normal film transport, replacement of the standard film gate with a double gate, and installation of a double prism in front of the lens so that left- and right-eye frames would be projected simultaneously.

Given that the purpose of single-strip 3-D systems was to simplify production and exhibition, it is easy to see why VeriVision did not succeed. Despite savings in film cost, VeriVision was a complicated system. It suffered serious light loss in projection, a defect of most single-strip 3-D systems where one lamp has to illuminate two frames of film.

John T. Rule—A Stereographic Pioneer

From 1936 to 1966, John T. Rule was a professor of engineering graphics at Massachusetts Institute of Technology. As a student of projective geometry, it was natural that he would acquire an interest in stereoscopic theory and the geometry of vision. He very quickly became an expert in stereoscopic drawing and transmission theory. With a series of three articles published in the *Journal of the Optical Society of America (JOSA)* from 1938 to 1941, Rule formulated elegantly practical techniques for drawing, photographing, and projecting both still and moving stereographic images.

Rule's interest in stereography led to some interesting work for him. He worked with Edwin H. Land on the development of the Vectograph and its use in stereoscopic photography. This experience proved fruitful for Rule during World War II when he created a special series of Vectographs for the U.S. Navy to teach basic navigation to military students. He was also in charge of the development of the Mark I machine gun trainer that used stereoscopic techniques to simulate combat conditions encountered in aerial warfare. Before the war, Rule had worked with John Norling on production of *In Tune with Tomorrow,* the dual 35-mm 3-D film created for the Chrysler Motor Corporation and shown in 1939 at the New York World's Fair.

These practical experiences in 3-D application gave Rule's stereoscopic theories real credibility. Along with Earle F. Watts, Rule also published two books, *Descriptive Geometry* and *Engineering Graphics.*[23] *Descriptive Geometry* contained an appendix on stereoscopic drawing that summarized Rule's previous article on the subject in *JOSA*. Rule frequently used polarizing 3-D slide projection in his classes to demonstrate geometric solids to his students.

Rule's first published article on the topic of 3-D, "Stereoscopic Drawings," in the August 1938 issue of *JOSA* was a comprehensive discussion of 3-D drawing that included both theory and practical techniques, as well as stereo pairs of his own creation. Rule propounded two distinct methods of stereoscopic drawing for viewing either by means of a Holmes-style stereoscope or a Wheatstone reflecting mirror system. "A perfect stereograph," wrote Rule, "would be one in which the resultant fused image appeared to be an exact duplicate of the original space scene in shape, size and location."[24]

This concern with orthoscopically correct stereo was worked out completely in Rule's second *JOSA* article, "The Shape of Stereoscopic Images," published in February 1941. Rule stated that the purpose of the paper was

to supply an exact tool for the analysis of stereoscopic effects so that it may become possible to determine the reasons for observable distortions and to predict the results of any proposed optical system. A further purpose is to furnish a base for the comparison of the psychological interpretation of images with their shapes on geometrical principles in order to open the way for the study of the degree of variation from orthostereoscopic images which is permissible in practical projection without annoyance.[25]

In the absence of a unified literature on the subject of stereoscopic transmission, Rule's articles in *JOSA* were foundational and important. His third *JOSA* article, "The Geometry of Stereoscopic Projection," dated April 1941, was Rule's attempt to create "a unifying article on the theory and practice" of stereoscopic projection. He acknowledged that "much of the material in this article is not new but only correlated into a consistent pattern." Here, the "little understood factors" that "cause the brain to misinterpret stereoscopic images in varying degrees" were discussed. They included the principles of convergence and accommodation and how they apply to parallel and nonparallel (converged) projection of the stereographic image. The important factor of stereoscopic magnification was also clearly explained as it affects the viewing experience.[26]

There is no doubt that John T. Rule was a champion of stereography. At the conclusion of his last *JOSA* article, he wrote: "No form of still photography in any way approaches the reality, nor yields the satisfaction, of properly photographed and projected natural color stereographs."[27]

Stereo Movies and the Avant-Garde

It should not be surprising that twentieth-century artists were interested in stereography and would attempt to make their own stereoscopic motion pictures. Science and technology were important concepts to the modernist artists of the early twentieth century. Marcel Duchamp, a premiere Dadaist and associate of the Surrealist artists, was very interested in stereo drawing and anaglyphs after he had found a copy of a 1912 French book titled *Les Anaglyphes geometriques* by H. Vuibert.[28] Primarily a scientific text about descriptive geometry, *Les Anaglyphes geometriques* was printed in red and green inks and included a lorgnette pair of spectacles for viewing the geometric shapes.

In 1918, while in Buenos Aires, Duchamp manipulated a stereoview card with a rather prosaic photograph of the ocean by pencil drawing an octohedron as a stereo pair of images on the card. In New York in 1920, working with fellow artist and photographer Man Ray, Duchamp made a stereoscopic motion picture, *Rotative Demisphere,* of his rotating motorized construction using two motion picture cameras. Man Ray was shooting stereoscopic still photographs at the time and made a stereoview of the physical setup showing the rotative glass plates. In his autobiography, published in 1963, Man Ray recalled the making of the stereoscopic motion picture with Duchamp:

> Then Duchamp came to me with projects: he had conceived an idea for making three-dimensional movies. Miss Dreier had presented him with a movie camera, and he obtained another cheap one—the idea was to join them with gears and a common axis so that a double, stereoscopic film could be made of a globe with a spiral painted on it.[29]

Unfortunately, Duchamp decided to develop the film himself. The process turned the film into "a mass of tangled seaweed. It had swelled and the film was stuck together." Saving some of the film, Duchamp and Man Ray examined two matching strips in an old stereopticon and saw the effect of relief. No capital was available to them at the time, however, so the project was abandoned. In 1968, near the end of his life, Duchamp was still experimenting with stereography by making a drawing with red and blue crayons that was titled *Cheminee anaglyphe (Anaglyphic chimney).* In 1927, when Man Ray was making *Emak Bakia,* a black-and-white film using stop-motion and double exposure of chess pieces and dice, 3-D filmmaking was very much on his mind. He wrote:

> As for being a purist to the extent of preferring old, silent, black and white film, this criticism is purely arbitrary because I insisted from the start on sound accompaniment, longed for the use of color and three-dimensions, even hoped for the addition of the sensations of warmth, cold, taste and smell to film, so that lastly the spectator, coming out in to the fresh air of the street, could be totally in enjoyment of all his senses, with the added advantage of being the principal actor![30]

Oskar Fischinger was a pioneering filmmaker of the twentieth century who created visual music on film with abstract paintings in motion that

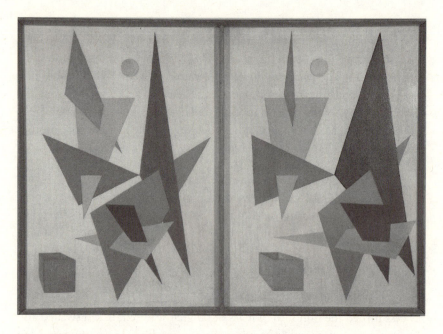

"Triangular Planes," stereo painting by Oskar Fischinger (1949) (Jack Rutberg Fine Arts, Los Angeles, California).

were accompanied with sound and music. Fischinger began to produce abstract films in 1922, and in a few years was synchronizing nonobjective imagery to popular records with a series he called *Studies*. These films were shown in theaters as advertisements for the recordings. Sixty years before MTV, they were the first music videos. Each of these studies ran three minutes and included approximately 5,000 drawings coordinated to the music. By 1935, Fischinger had made color films with works titled *Circles* and *Composition in Blue*.

In 1947, after immigrating to America from Germany to work for Paramount Studios and Walt Disney, Fischinger began to create stereo paintings. These were side-by-side oil paintings on masonite panels, usually hinged together. After completing the film titled *Motion Painting I* in 1947, Fischinger launched into his experiments in stereo painting. William Moritz, Fischinger's biographer, wrote that "he worked first analyzing a set of old stereoscope slides and pairs of photos he took of himself. Then he carefully prepared a dozen canvases in which complex and varied types of abstract shapes and surfaces are seen from right-eye and left-eye perspectives in parallel panels."[31]

"Stereo Film," stereo pairs from film by Oskar Fischinger (1952) (Jack Rutberg Fine Arts, Los Angeles, California).

After producing a number of stereo paintings over a four-year period, Fischinger felt prepared enough to produce the half-minute *Stereo Film* in a dual 35-mm format to be projected on a silver screen and viewed through polarizing filters. "The technique is basically that of *Motion Painting I*," noted Moritz, "except that Fischinger had built a special apparatus to accommodate two side-by-side paintings, and by now he could freely paint from his head 3-D images—this square would sit here, and this here—for he had completely mastered the mathematics, the formulas, the calligraphy of depth." Moritz observed that Fischinger's short 3-D film vividly depicted a "succession of a dozen colored squares which form a perfect and astounding perspective alignment from near the viewer to the far distance."[32]

Unfortunately, the artist had no luck in seeking completion funds from various foundations and prospective backers, and the project was abandoned. The thirty-second stereoscopic film is today periodically screened in 3-D—most recently in dual 16 mm at the Jack Rutberg Gallery in Los Angeles on May 2, 2000, and in dual 35 mm at the first World 3D Expo in Hollywood on September 20, 2003. In 1953, Fischinger had a one-man exhibit of his paintings at the San Francisco Museum of Modern Art. Prominently placed on the title wall of the exhibit were three stereo paintings that greeted patrons as they entered the museum.

The films and paintings of Oskar Fischinger have had a tremendous influence on modern artists. "Fischinger was one of the great formative influences in my life," declared 3-D filmmaker Norman McLaren: "In the early 1950s, I had the opportunity and great pleasure of visiting Fischinger and his wife at their home in California. . . . I discovered that Oskar was interested not just in filmmaking, but was into all kinds of other experiments, the most intriguing of which for me was his stereoscopic paintings, for I myself had been dabbling in binocular drawings." McLaren more than dabbled in stereography. He made the pioneering dual 35-mm animated 3-D films *Around Is Around* and *Now Is the Time (To Put on Your Glasses)* in 1951 for the Festival of Britain. After visiting with Fischinger, McLaren said, "I felt I had met someone with a truly inventive and exploratory spirit, and an artist who had pioneered a new path in the history of cinema."[33]

The Festival of Britain

The Festival of Britain in 1951 was a presentation of Britain's accomplishments in science and art, a century after the Great Exhibition of 1851. On the South Bank of the Thames River in London, the Telecinema, a futuristic theater seating 400 and designed by Wells Coates, was constructed to show live, large-screen television and stereoscopic motion pictures.

The Telecinema and its program of four stereoscopic films, as well as documentaries and television shows, was one of the outstanding successes of the festival. It grossed $225,000, a small fortune in postwar Britain, for seven to nine shows a day over the twenty-two weeks of the festival, and was attended by nearly half a million patrons.

The Magic Box, a feature film in Technicolor produced by Michael Balcon of Ealing Studios, told the story of film pioneer William Friese-Greene and was made specifically for the Festival of Britain. Englishman Raymond Spottiswoode was appointed technical director for production of stereoscopic movies, and the National Film Board of Canada was invited to contribute an animated 3-D film to be created by Norman McLaren.

All the 3-D motion pictures at the Telecinema were produced and projected in dual 35-mm interlock, and these were the first stereoscopic motion pictures to be projected in stereophonic sound. Special loudspeaker units and four-channel magnetic stereophonic playback for the binaural music reproduction was built by the British Thomson-Houston Company (B.T.-H.) which also designed a new type of screen surround providing a "picture-frame" effect of variable light intensity. A silver screen provided by Stableford Screens was used to view the twin-strip stereo projected with Polaroid filters, and the audience was equipped with polarizing 3-D glasses. B.T.-H. also provided the two projectors that were interlocked with selsyn motors to run in sync.

For stereoscopic motion picture photography, Spottiswoode hired Leslie Dudley, who built two twin camera rigs using dual 35-mm Newman-Sinclair cameras facing inward on a baseplate toward two mirrors at 45-degree angles so that the stereobase could be varied from 1 to 8 inches with a half angle of convergence (stereoangle) from 0 to 5 degrees. This unit was used for stereoscopic photography in the black-and-white film *A Solid Explanation,* a humorous illustration of the principles of stereoscopy intercut with scenes of various animals (notably the giraffe) shot at the London Zoo. Raymond Spottiswoode described *A Solid Explanation* thus:

> This film is built round the character of an eminent professor who believes that an audience cannot appreciate a three-dimensional film unless it has first grasped the principles of stereoscopic transmission. (Any resemblance to the present writer is wholly coincidental!) While he becomes more and more mixed up in tangled phrases and demonstrations which don't come off, the camera cuts away to sequences which clearly show the heightened reality of the three-dimensional film.[34]

As a result of the experience of producing stereoscopic films for the Festival of Britain, Raymond Spottiswoode, along with his brother Nigel, set about evolving a comprehensive theory of stereoscopic transmission, which was published as a book under that title in 1953.[35] The Spottiswoode brothers claimed to have discovered that "a single master equation determines the shape of the [stereoscopic] image under all possible variations in the camera and projection systems."[36]

For stereoscopic photography of a film in color, Spottiswoode hoped to use the new single-strip Eastman color Monopack negative. After testing, it was found to be unsatisfactory, so Technicolor Ltd. of England offered two of the three-strip Technicolor cameras mounted alongside each other on a baseplate with a tapered wedge for a variable angle of convergence. A selsyn system was used for interlock-sync and follow-focus of the cameras. The stereobase with two of the bulky three-strip Technicolor cameras side by side was 9½ inches. "This suggested that the film should be composed mainly of long shots," wrote Spottiswoode, "in which what we call extra-stereoscopic factors—perspective, masking, light and shade and so on—should as far as possible counteract the miniaturizing effect produced by the exaggerated stereobase."[37]

Filming was done from a platform moving slowly down the river for production of the initial version of *The Distant Thames,* subsequently expanded and retitled *Royal River.* Polled upon exiting the 3-D films, 52 percent of the audience picked *The Distant Thames* as their favorite. When questioned, "Did you notice anything peculiar about the film?" regarding the miniaturization effect, 62 percent of the audience said they did not notice the effect.

The program of stereoscopic films began with McLaren's *Now Is the Time (To Put on Your Glasses)* about which reviewer Norman Jenkins wrote, "There is nothing quite like this in the average cinemagoer's experience," and that it has "the added unfamiliarity of synthetic sound—photographed patterns."[38] The sound track for this film, to use McLaren's term, was a "stereophonic animated sound track." McLaren wrote an invaluable paper describing the production of both *Now Is the Time* and *Around Is Around.* To make both these films, McLaren had "to synthesize three-dimensional space, from two-dimensional subject matter."[39]

A conventional camera, optical setup, and animation stand was used for *Now Is the Time.* Parallax for left- and right-eye views was created with movable cutouts in the artwork. The cutouts were moved left or right according to calculations based on screen size, which, at the Telecinema, was 15 feet. McLaren consulted John T. Rule's classic paper on stereoscopic projection, as well as the work of Spottiswoode on the subject. Parallax by lens shift was also created with optical prints made from the hand-drawn negative to produce left- and right-eye optical negatives. Optical prints made from the hand-drawn negatives were used to produce left- and right-eye optical negatives. Color separation negatives (yellow, cyan, and magenta) were made from the combined elements for release color printing by Technicolor in England.

The synthetic sound for *Now Is the Time* was produced by photographing patterns of black-and-white sound waveforms on to the sound track area of 35-mm film using standard animation techniques. Lou Applebaum composed a special stereophonic musical score for *Around Is Around,* also creating a four-channel stereophonic recording that made use of different instruments and the spatial location in the auditorium of the Telecinema.

Around Is Around used both standard cel animation double-punch techniques and frame-stagger photography of oscillograph patterns in motion. The oscillograph patterns were controlled manually with a knob, and the Bell and Howell camera ran at eight to twelve frames per second for greater control of pattern modulation. The black-and-white stereo pairs were optically printed as yellow, cyan, and magenta color separation negatives for release printing in Technicolor.

Gerald Pratley, reviewing McLaren's stereoscopic creations for *Films in Review,* called *Around Is Around* "a ballet of lines, ever-changing in circular motion," and gave it the highest praise:

> McLaren's second film, *Around Is Around,* is the most perfect and beautiful of all his remarkable attempts to use celluloid as the artist uses brush and canvas and the poet pen and paper. For the first time in motion pictures we see around a spherical shape, instead of seeing it flat. This alone is an exciting revelation.[40]

The Festival of Britain marked an end to the novelty period of stereoscopic cinema. The century between the Great Exhibition of 1851 and the Festival of Britain was an extended laboratory of aspiration and innovation for stereoscopic cinema, with many failures and a few artistic successes. But with the stereographic animation of Norman McLaren, there was no question that 3-D motion pictures had at last achieved the status of art.

Promotional art showing Polly Bergen in MGM's 3-D feature film *Arena* (1953) emphasizes the emergence effect (Ray Zone Collection).

Epilogue

> When all film is stereoscopic, and we have forgotten that we ever
> accepted the convention of the flat-image as real, it seems unlikely
> that we shall remark on the stereoscopic film's appearance of real-
> ity, any more than at present we remark on the conventional flat-
> ness of the two-dimensional film.
> —Ivor Montague, "The Third Dimension: Film of the Future?"

DID THE FESTIVAL OF BRITAIN play a part in the 1950s 3-D boom in
Hollywood? R. M. Hayes certainly thinks so. He wrote that "the influence
of the film presentations in the Telecinema would have a worldwide effect,
for it was here in May 1951 that dual projection 3-D in full color with
interlock stereophonic sound was to be realized. From this was to evolve
the boom years of 1953 and 1954, regardless of claims otherwise."[1]

Raymond Spottiswoode expressed his wish that the Telecinema would
continue: "We are hopeful that the Telecinema will remain in existence
under the progressive management of the British Film Institute as a place
where three-dimensional films and live television can continue to foreshadow the entertainment of the future." And shortly after the Telecinema
had completed its run in the fall of 1951, Spottiswoode posed the provoca-
tive question, "If these little films, made on a budget of a few thousand
pounds, could attract such enormous audiences, and cause an audible thrill
to run through the house at each performance, what would not be the
stimulating effect on the box office of three-dimensional films made with
all the resources of Hollywood?"[2]

A year later, after *Bwana Devil* opened on November 26, 1952, Spottiswoode would have his answer. But Spottiswoode, with his brother Nigel

and others, did not wait for that day. They formed Stereo Techniques Ltd. and commenced production on eighteen short stereoscopic motion pictures, released in the United Kingdom from 1952 to 1955. Stereo Techniques Ltd. also hired the Richardson Company in Hollywood to build the Space Master stereo cinema camera rig, using two Cameraflexes, and shot a 3-D feature film, *The Diamond Wizard,* with it.[3]

Hayes also contended that it was Raymond Spottiswoode who "created the term '3-D,' not the Hollywood tradepaper *Variety,* which has always taken and been given credit for such."[4] Spottiswoode's first use of the term may have been in the seminal paper "Basic Principles of the Three-Dimensional Film," published in the October 1952 issue of the *Journal of the Society of Motion Picture and Television Engineers,* which he coauthored with Nigel Spottiswoode and Charles Smith.[5]

This extended discussion of the principles of stereoscopy was a precursor to the Spottiswoode brothers' book, *The Theory of Stereoscopic Transmission and Its Application to the Motion Picture,* which the University of California Press subsequently published in 1953, while the 3-D frenzy was taking place in Hollywood. In the acknowledgments to that book, the Spottiswoodes expressed their gratitude to Denis Forman, director of the British Film Institute (BFI), and Gerald Barry, general director of the Festival of Britain, "for the opportunity they gave us alike in the field of pure thought and of practical experience." They also stated that they felt their book "could not have come about within the ambit of the motion picture industry, organized as it is at present either in Britain or America."[6] The Spottiswoodes had been to Hollywood and had seen *Bwana Devil* prior to publication of their book, and they refer to the film as "a veritable casebook of stereoscopic errors."[7]

By this time, the flashpoint that started the Hollywood 3-D boom had already taken place. What had been going on in Hollywood during the Festival of Britain? Early in 1951, a pair of Hollywood cameramen, Lothrop Worth and Friend Baker, had begun to experiment with 3-D. Lothrop Worth enjoyed a fifty-year career in motion pictures as a cameraman. Working with Friend Baker in 1951, Worth built a 16-mm prototype for a lens-to-lens, dual-camera, 3-D system. When screenwriter Milton Gunzberg saw the 3-D footage that Worth had shot with the camera, he persuaded his brother Julian, a Beverly Hills ophthalmologist, to invest in a 35-mm dual-camera 3-D rig. A letter from Friend Baker to Arnold Kotis, editor of the *Journal of the Society of Motion Picture and Television Engineers,* documented this seminal moment. The letter, in the possession of the author, was dated June 23, 1951, and typed on Baker and Worth letterhead. Baker began: "I am sorry to be so late in answering your letter of last month but it is only of the last few days that we had decided to do anything about publicity on the Stereo unit." The Festival of Britain had commenced in April 1951, and, though he doesn't mention it, there is a good possibility that Baker was aware of it. The telling item in Baker's letter was a handwritten postscript: "P.S. Recently completed equipment for 35mm motion picture which gives 16 foot stereo in color."

Worth and Baker, working with O. S. Bud Bryhn, built the 35-mm dual-camera rig with micrometer and mirrors, which ultimately came to be known as Natural Vision, so named by Julian Gunzberg for its convergence feature, which mimicked the human eyes. In a 1999 interview with film historian Mike Hyatt, Worth discussed promoting stereoscopic motion pictures to the Hollywood community:

> We made a lot of camera tests. We even made a test for the press. And we had got a projection deal with a little studio in Culver City. We rented it and we invited all the press. We had some sensational footage. And all they wanted to know was "Who's going to make a picture and when?" We couldn't seem to get anybody. We had critics come in who would say, "You can't shoot close-ups." So we'd shoot a lot of close-ups. "You can't shoot running shots with it," they would say. We'd run a car and shoot running shots. This went on for a long time.[8]

Taking Worth's recollection into account, reports of the success of 3-D films at the Festival of Britain did not seem to affect the Hollywood community. Andrew Dowdy, in his book, *The Films of the Fifties,* confirms the situation:

> Earlier that year (1951) Milton Gunzburg and his brother Julian formed the Natural Vision Corporation, but their attempt to sell the major studios on their stereoscopic process wasn't successful. Columbia turned them down. At Fox, Darryl F. Zanuck said nothing requiring glasses could succeed. . . . Paramount also turned down the Gunzburgs. MGM, where John Arnold, chief of the camera department, had worked privately with 3-D for thirty-five years, took an option but allowed it to lapse.[9]

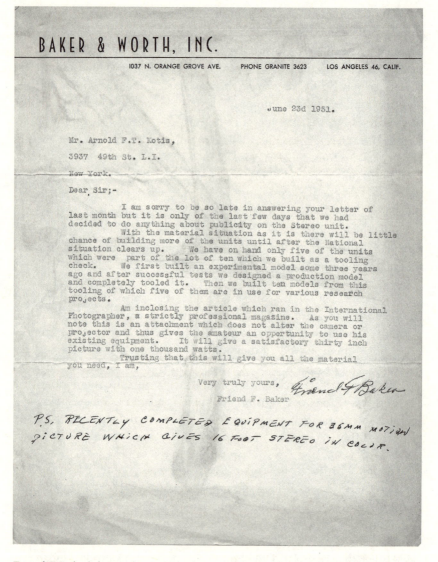

Friend F. Baker's letter, documenting a seminal moment in 3-D film history (1951).

Independent producer Arch Oboler was eager to make a 3-D film. Using $10,000 of his own money, Oboler cut a deal with the Gunzbergs and began production with Natural Vision for filming of *Bwana Devil* in June 1952, on location in the Malibu foothills, with Lothrop Worth as camera operator and Joe Biroc as director of photography. Within two days

The Natural Vision stereo camera that filmed *Bwana Devil* used two Mitchell cameras lens to lens with 45-degree mirrors (1952) (Ray Zone Collection).

of *Bwana Devil*'s November 26 release, Jack Warner licensed the Natural Vision process from the Gunzbergs and signed Lothrop Worth to shoot *House of Wax* (1953) in 3-D. *Bwana Devil* was an immediate success, grossing more than $95,000 in the first week in two theaters; by December 1952, it was breaking movie attendance records.[10]

It seems very likely that the promise of a financial windfall, rather than stereoscopic delight, was the motivating factor for Hollywood to initiate the 3-D boom of the 1950s. For his part, Edwin H. Land had always been a proponent of the stereoscopic experience, whether it was with motion pictures, Vectographs, or holograms. Polaroid stock rose 30 percent in early 1953 as the company granted Natural Vision an exclusive one-year contract. At ten cents each, the Gunzbergs sold 100 million 3-D glasses to exhibitors.

Despite the frenzied 3-D activity, Land encouraged the motion picture industry to respect stereography:

Now the talents of this same industry have, quite literally, a new dimension to exploit; all the space in the world (again, quite literally) to work in. This space can be thrown away, as a passing nov-

MASSACHUSETTS INSTITUTE OF TECHNOLOGY
CAMBRIDGE 39, MASSACHUSETTS

COURSE IX
General Science &
General Engineering

March 25th, 1952

Mr. J. A. Norling,
245 West 55th Street,
New York 17, N. Y.

Dear Jack:

   I have been over your definitions. Some of them I think should be changed. In order they are:

Disparate Images: I don't like your use of the word "image". To avoid confusion we should carefully distinguish between three things:-

 1. The stereoscopic image which is the solid image in space seen by the observer. It exists only to the observer.

 2. The screen pair or pictures which should not be called images at all.

 3. The film pair or pictures which again should not be called images.

An image is an optical phenomenon not a picture.

Focal Distance: I don't like the wording. I think it should read: "The distance from the principal node of the camera lens to the film. Except for extremely close views the stated focal length of the lens is a sufficiently close approximation."

Horizontal Error: I don't understand what you mean by Horizontal Error. A "spacing apart" is always there by choice. It might read: "Any error that causes a different spacing than that desired between homologous points." The error would be there whether it caused eyestrain or not.

Hyperstereoscopy and Hypostereoscopy: are much too loose. I would define a hyperstereoscopic image as one whose three-dimensional image size is smaller than the actual object in space (see toward the end of this letter). Some people tend to define the hypostereoscopic image as one that has exactly the same shape but is smaller than the object, excluding the image in which the depth is exaggerated with respect to the width and height. Some clarification is needed here.

Indicated Interaxial: I don't think this is "established by the distance of the object and the focal length of the lens employed.". I would make the definition read: "The interaxial necessary to obtain some desired effect from the given set of conditions." The desired effect might be either hyperstereoscopic or orthostereoscopic.

John T. Rule letter to John Norling, a meeting of stereographic minds (1952).

elty. Or the industry can determine to use the space, the solid realism of the new dimension, as a true addition to the medium they have created. This would be our best assurance that people will be wearing glasses after the novelty has worn off; they will be enjoying themselves, carried away by the realism of the art.[11]

- **Vertical Displacement:** Question is where? Should read: "In projection: When the horizontal axis of one member is higher or lower than that of the other on the screen."

- **Vertical Error:** Add: "on the screen".

     I think we will find as we go along that many of these definitions are too loose and will lead to misunderstandings which will need to be cleared up as they appear. Those which try to establish limits to what is in fact a continuum will have to have precisely-defined arbitrary limits. An example is **Maximum Viewing Angle.** The definition as given is completely meaningless until some criteria specifying what degree of "marked consciousness of shape distortion" shall be arbitrarily taken as a limit.

     There are other words, of course, which should be defined. Examples are: giantism, cardboarding, keystoning, differential verticals, orthostereoscopic, pseudostereoscopic, and many others.

- **Orthostereoscopic Image:** An image having exactly the same size and shape as the original object.

- **Hyperstereoscopic Image:** An image smaller in size but similar in shape to the original object.

- **Hypostereoscopic Image:** An image larger in size but similar in shape to the original object.

- **Elongated Image:** An image which, with respect to the original object, is exaggerated in depth compared to width and height.

- **Cardboarded Image:** An image which with respect to the original object is decreased in depth compared to the width and height.

- **Horizontal Keystoning:** Projecting in such a way that equal horizontal distances at the top and bottom of the film are unequal on the screen.

- **Vertical Keystoning:** Projecting in such a way that equal vertical distances on the left and right of the film are unequal on the screen.

- **Differential Vertical:** When either or both pictures are keystoned the vertical displacement of the screen pictures differs across the screen. This is not the same as verticals caused by rotation.

     I will be in New York Thursday, 3rd of April, and will drop in on you some time in the morning. This is rather a hasty going-over but will indicate a desire on my part for more precision.

Sincerely,

John T. Rule

co - SMPTE No.417

Page two of Rule's letter to Norling shows Norling's meticulous checkmarks for each item.

     In March 1952, seven months before the release of *Bwana Devil,* John T. Rule and John Norling were corresponding with the aim of establishing a stereoscopic glossary and technical standards for 3-D motion pictures.[12] A year later, an SMPTE Committee on Stereo Motion Pictures was active, with

Norling serving as chairman. The Stereo Committee also included John T. Rule, Raymond Spottiswoode, and Clarence Kennedy as members.[13]

Spottiswoode submitted a drawing for a method of alignment for 35-mm twin-strip stereoscopic projection to the SMPTE Stereo Committee. The individual left- and right-eye strips were identified separately with dual leader that also served as a registration guide on screen.[14] A subsequent memo to the committee noted that "Mr. Spottiswoode deserves the thanks of the Society for his excellent job, and as Prof. Rule says, '—is to be congratulated on his precision, clarity and firm grasp of the essentials.'"[15]

Despite these efforts of the best stereographic minds to give 3-D motion pictures a permanent place in exhibition, by 1954 the Hollywood 3-D movie boom was winding down. Did the novelty factor of stereoscopic motion pictures, the gimmick of off-the-screen effects, relegate it, once again, to a short life in exhibition? Film historian William Paul has argued:

> The chief problem with 3-D as it was applied to conventional Hollywood stylistic practices, and most especially to prestigious films, lies in what continues to be its primary attraction for audiences, and the thing that most fully distinguished it from the myriad new screen systems of the 1950s: the phenomenon of "negative parallax," or the emergence effect. So long as the emergence effect remained central to the experience of 3-D, the process inevitably became tied, as I have argued elsewhere, to exploitation fare.[16]

To some extent, this is a stereographic philosophy with which the Spottiswoode brothers would have agreed. In 1952, Stereo Techniques Ltd. produced a ballet film, *The Black Swan,* which had no off-the-screen effects. In fact, the stereo window itself, the periphery of the onscreen image, was made to come out into the audience space to eliminate the emergence effect. Nevertheless, after making *The Black Swan,* the Spottiswoode brothers wrote:

> It is this contrast between the familiar movie theater, associated with tame flat films, and the seemingly fierce and powerful things which can be made to inhabit it, which creates gasps and thrills unknown to the cinema since its earliest infancy. Obviously this is a stage which will pass, and the image in space will come to be accepted as a thing as humdrum as the flat films which took an earlier generation's breath away.[17]

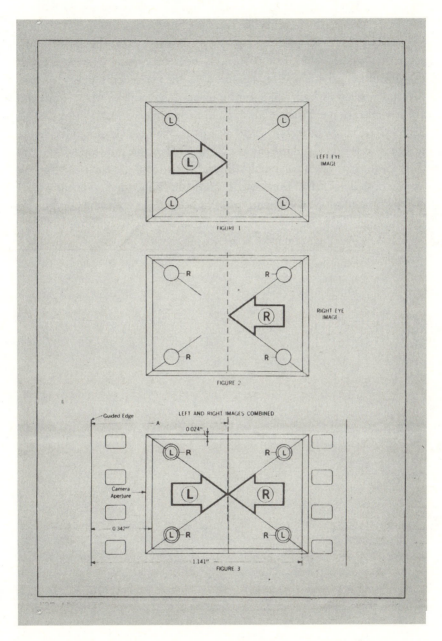

Raymond Spottiswoode's proposed standard for 35-mm twin-strip alignment (1952).

In the twenty-first century, stereoscopic filmmakers continue to grapple with the aesthetic issues of telling a story in the z-axis. The use of visual space that is both behind and in front of the motion picture screen expands the possibilities for narrative elaboration. This expanded narrative palette presents not a drawback but a luminous new challenge to which filmmakers must rise. Sergei Eisenstein was filled with a sense of the possibilities for 3-D movies when he wrote: "Will all this not call for absolutely new arts, unheard-of forms and dimensions ranging far beyond the scope of the traditional theatre, traditional sculpture and traditional . . . cinema ?"[18]

As digital 3-D cinema evolves, simplified production and exhibition of stereoscopic motion pictures promises to become an enduring component of the twenty-first-century entertainment landscape. The alluring possibility of a new visual grammar for stereoscopic motion pictures presents itself once again.

Notes

1. Stereography Begins

1. Euclid, *Optica,* in *Quae Supersunt Omnia,* ed. David Gregory (Oxford: n.p., 1703), 619–20.

2. Brian Bowers, *Sir Charles Wheatstone* (New York: Crown, 1975), 192.

3. *Galen on the Usefulness of the Parts of the Body (De usu partium),* trans. with commentary by Margaret Tallmadge May (Ithaca, N.Y.: Cornell University Press, 1968), 495–96.

4. Leonardo da Vinci, *The Art of Painting,* ed. Alfred Werner (New York: Philosophical Library, 1957), 82. After making this statement, da Vinci launches into a complex explanation accompanying schematic drawings of the eyes viewing a sphere, which editor Alfred Werner interprets by saying that, in painting, "Leonardo objects to the use of both eyes . . . which renders the perspective false in the painting" (83).

5. Herbert Mayo, *Outlines of Human Physiology,* 3rd ed. (London: Burgess and Hill, 1833), 288.

6. Charles Wheatstone, "Contributions to the Physiology of Vision, Part the First: On Some Remarkable, and Hitherto Unobserved, Phenomena of Binocular Vision," *Philosophical Transactions of the Royal Society of London* 4 (June 21, 1838): 76–77.

7. Charles Wheatstone, "Contributions to the Physiology of Vision, Part the Second: On Some Remarkable, and Hitherto Unobserved, Phenomena of Binocular Vision," *Philosophical Transactions of the Royal Society of London* 142 (January 15, 1852): 6–7.

8. Ibid., 7.

9. Nicholas J. Wade, ed., *Brewster and Wheatstone on Vision* (London: Academic Press, 1983), iii.

10. Sir David Brewster, *The Stereoscope: Its History, Theory, and Construction* (London: John Murray, 1856), 19.

11. Ibid., 20.

12. Ibid., 21.

13. Oliver Wendell Holmes, "The Stereoscope and the Stereograph," *Atlantic Monthly* 3 (June 1859), at http://www.stereoscopy.com/library/holmes-stereoscope-stereograph.html (accessed March 21, 2006).

14. Ibid.

15. Ibid.

16. Oliver Wendell Holmes, "Sun-Painting and Sun-Sculpture: With a Stereoscopic Trip across the Atlantic," *Atlantic Monthly* 8, no. 45 (1861), at http://www.yale.edu/amstud/infoev/stereo.html (accessed March 21, 2006).

17. Oliver Wendell Holmes, *Philadelphia Photographer* (January 1869), at http://www.yale.edu/amstud/infoev/stereo.html (accessed March 21, 2006).

18. Holmes, "Stereoscope and Stereograph."

19. Ibid.

20. William Culp Darrah, *Stereo Views: A History of Stereographs in America and Their Collection* (Gettysburg, Pa.: Times and News, 1964), 8.

21. Reese V. Jenkins, *Images and Enterprise, Technology and the American Photographic Industry, 1839 to 1925* (Baltimore: Johns Hopkins University Press, 1975), 50, 49.

22. Laura Burd Schiavo, "From Phantom Image to Perfect Vision: Physiological Optics, Commercial Photography, and the Popularization of the Stereoscope," in *New Media, 1740–1915,* ed. Lisa Gitelman and Geoffrey B. Pingree (Cambridge: MIT Press, 2003), 119.

23. Ibid., 120.

24. *Anthony's Photographic Bulletin* (November 1, 1870), 208, as cited in ibid.

25. William Darrah, *The World of Stereographs* (Gettysburg, Pa.: W. C. Darrah, 1977), v.

26. Annette Michelson, *The Art of Moving Shadows* (Washington, D.C.: National Gallery of Art, 1989), 39.

27. John Fell, *Film and the Narrative Tradition* (Norman: University of Oklahoma Press, 1974), xv.

28. Published as H. Mark Gosser, *Selected Attempts at Stereoscopic Moving Pictures and Their Relationship to the Development of Motion Picture Technology, 1852–1903* (New York: Arno Press, 1977).

29. Ibid., 2.

30. Laurent Mannoni, "The 'Feeling of Life': The Birth of Stereoscopic Film," in *Paris in 3D,* ed. Francoise Reynaud, Catherine Tambrun, and Kim Timby (Paris: Booth-Clibborn, 2000), 136.

2. The Peep Show Tradition

1. Richard Balzer, *Peepshows: A Visual History* (New York: Harry N. Abrams, 1998), 20.

2. Olive Cook, *Movement in Two Dimensions* (London: Hutchison, 1963), 23.

3. Charles Baudelaire, "The Modern Public and Photography," in *Classic Essays on Photography*, ed. Alan Trachtenberg (New Haven, Conn.: Leete's Island Books, 1980), 87.

4. Martin Quigley Jr., *Magic Shadows: The Story of the Origin of Motion Pictures* (New York: Quigley, 1960), 111.

5. C. W. Ceram, *Archaeology of the Cinema* (New York: Harcourt, Brace and World, n.d.), 60–61.

6. Paul Wing, *Stereoscopes: The First Hundred Years* (Nashua: Transition, 1996), 55.

7. Paul Wing, letter to the author, November 22, 2000.

8. David W. Wells, *The Stereoscope in Ophthalmology with Special Reference to the Treatment of Heterophoria and Heterotropia* (Boston: E. F. Mahady, 1928), ii.

9. Joseph Plateau, "Troisieme note sur de nouvelles applications curieuses de la persistence des impressions de la retine," *Bulletin de l'Academie royale des sciences de Bruxelles* 16, no. 7 (1849): 38–39.

10. "Bulletin de la Correspondance," *La Lumiere* 2 (May 22, 1852): 88.

11. *Le Moniteur de la photographie* 15 (October 15, 1865): 114–16.

12. Antoine Claudet, "Moving Photographic Figures," *Photographic Journal* 9–10 (September 15, 1865): 143. Some of the publications that reprinted this article include *Journal of the Franklin Institute* (November 1865), *British Journal of Photography* (no. 280, 1865), *Photographic Journal* (September 15, 1865), *Philosophical Magazine* (October 1865), and *Bulletin de la Societe Francaise de Photographie* (no. 6, 1865).

13. Jules Duboscq, "Sur le Stereoscope," *Bulletin de la Societe Francaise de Photographie* 3 (March 1857): 77–78.

14. Quoted in H. Mark Gosser, *Selected Attempts at Stereoscopic Moving Pictures and Their Relationship to the Development of Motion Picture Technology, 1852–1903* (New York: Arno Press, 1977), 88–89.

15. Laurent Mannoni, "The 'Feeling of Life': The Birth of Stereoscopic Film," in *Paris in 3D,* ed. Francoise Reynaud, Catherine Tambrun, and Kim Timby (Paris: Booth-Clibborn, 2000), 136.

16. Quoted in Gosser, *Selected Attempts at Stereoscopic Moving Pictures,* 88.

17. Quoted in ibid., 89.

18. Ibid., 89–90.

19. Ibid., 106.

20. Philippe Benoist, "Instrument d'optique dit images animees," French Patent No. 16,055 (April 5, 1853); "An Improvement in the Construction of Sterioscopes [*sic*]," British Patent No. 1,965 (August 23, 1856).

21. Adam Jundzill, "An Instrument for Animating Stereoscopic Figures," British Patent No. 1,245 (May 24, 1856).

22. Victor Pierre Sies, "Animated Stereoscope," French Patent No. 43,297 (December 19, 1859), as cited in Mannoni, "Feeling of Life," 142.

23. Fume fils and H. Tournier, "Stereoscope employe comme phenakistiscope," French Patent No. 46,340 (August 14, 1860).

24. Andre David, "Appareil binoculaire dit stereoscope anime," French Patent No. 100,076 (August 22, 1873).

25. Henry DuMont, French Patent No. 42,843 (November 17, 1859), and Belgian Patent No. 7,321 (March 26, 1859).

26. Gosser, *Selected Attempts at Stereoscopic Moving Pictures,* 93.

27. Henry DuMont, "Appareil cylindrique pour obtenir rapidement douze clichés successives," *Bulletin de la Societe Francaise de Photographie* 8 (February 1862): 34–35, translated by Gosser, *Selected Attempts at Stereoscopic Moving Pictures.*

28. Quigley, *Magic Shadows,* 109.

29. Gosser, *Selected Attempts at Stereoscopic Moving Pictures,* 107.

30. Ibid., 98 n.30. The article cited is "Letter to Publisher," *Photographic News* 4 (May 11, 1860): 20.

31. Quigley, *Magic Shadows,* 111.

32. C. Sellers, "Exhibiting Stereoscopic Pictures of Moving Objects," U.S. Patent No. 31,357 (February 5, 1861), 2.

33. Stephen Herbert, "Re: On the Motoroscope," email to the author, March 23, 2005.

34. James Laing, "On the Motoroscope," *Proceedings of the Royal Scottish Society of Arts* (March 14, 1864): 1.

35. Ibid, 2.

36. Ibid.

3. Motion Pictures Begin

1. Homer Croy, *How Motion Pictures Are Made* (New York: Harper and Brothers, 1918), 8.

2. Terry Ramsaye, *A Million and One Nights: A History of the Motion Picture* (New York: Simon and Schuster, 1926), 21.

3. Leslie Wood, *The Miracle of the Movies* (London: Burke, 1947), 66.

4. Rebecca Solnit, *River of Shadows: Eadweard Muybridge and the Technological Wild West* (New York: Viking, 2003), 228.

5. Gordon Hendricks, *Eadweard Muybridge: The Father of the Motion Picture* (London: Secker and Warburg, 1975).

6. Solnit, *River of Shadows,* 80.

7. Ibid., 41.

8. Ibid., 53.

9. Ibid., 158.

10. H. Mark Gosser, *Selected Attempts at Stereoscopic Moving Pictures and Their Relationship to the Development of Motion Picture Technology, 1852–1903* (New York: Arno Press, 1977), 168. Gosser cites Eadweard Muybridge, *Animals in Motion* (London: Chapman and Hall, 1899), 2–3.

11. Stephen Herbert, ed., *Eadweard Muybridge: The Kingston Museum Bequest* (East Sussex: Projection Box, 2004), 14.

12. Quoted in Solnit, *River of Shadows,* 195–96, 200.

13. Etienne-Jules Marey, "Locomotion de l'homme: Images stereoscopiques des trajectories que drecrit dans l'espace un point du tronc pendant la marche, la course et les autres allures," *Les Comptes Rendus de l'Academie des Sciences* 100, no. 22 (June 2, 1885): 1359–63, translated by Laurent Mannoni, "The 'Feeling of Life': The Birth of Stereoscopic Film," in *Paris in 3D,* ed. Francoise Reynaud, Catherine Tambrun, and Kim Timby (Paris: Booth-Clibborn, 2000), 139.

14. Mannoni, "Feeling of Life," 139.

15. Etienne-Jules Marey, *Le Mouvement* (New York: D. Appleton, 1895), 21–22.

16. Quoted in Gosser, *Selected Attempts at Stereoscopic Moving Pictures,* 171.

17. A "caveat," as noted in *Webster's Dictionary* of 1888, is "a description of any new invention or discovery, lodged in the office before the patent right is taken out." This attribution is by Gordon Hendricks, *The Edison Motion Picture Myth* (Berkeley: University of California Press, 1961), 15.

18. Thomas A. Edison, "Motion Picture Caveat I," U.S. Patent Office (October 17, 1888).

19. Thomas A. Edison, "Motion Picture Caveat IV," U.S. Patent Office (November 2, 1889).

20. William K.-L. Dickson and Antonia Dickson, *History of the Kinetograph, Kinetoscope and Kineto-Phonograph* (New York: Albert Bunn, 1895), 19.

21. Ibid., 16–18.

22. Edison Letter Book, 6/13/91–8/24/91, quoted in Hendricks, *Edison Motion Picture Myth,* 127.

23. Hendricks, *Edison Motion Picture Myth,* 68, 71.

24. T. A. Edison, "Apparatus for Exhibiting Photographs of Moving Objects," U.S. Patent No. 493,426 (March 14, 1893), 3.

25. Ibid., 2.

26. As cited in Gosser, *Selected Attempts at Stereoscopic Moving Pictures,* 262.

27. Quoted in Hendricks, *Edison Motion Picture Myth,* 136.

28. Willard Steward and Ellis Frost, "Kinetosope," U.S. Patent No. 588,916 (August 24, 1897), 3.

29. E. W. Scripture, "A Method of Stereoscopic Projection," *Scientific American* 73 (November 23, 1895): 32.

30. Gosser, *Selected Attempts at Stereoscopic Moving Pictures,* 265.

31. *Electrical Review* (Great Britain) (November 1881), 96–97, cited in Hendricks, *Edison Motion Picture Myth,* 94.

32. Hendricks, *Edison Motion Picture Myth,* 142.

33. Mannoni, "Feeling of Life," 139.

34. Georges Demeny, "Modified Phonoscope Device," British Patent No. 12,794 (June 30, 1893), 3.

35. E. J. Wall, "Stereoscopic Cinematography," *Transactions of the Society of Motion Picture Engineers* 10 (February 1927): 328.

36. Gosser, *Selected Attempts at Stereoscopic Moving Pictures,* 201.

37. Gordon Hendricks, *Beginnings of the Biograph: The Story of the Invention of the Mutoscope and the Biograph and Their Supplying Camera* (New York: Beginnings of the American Film, 1964), 2.

38. F. Moniot and L. Garcin, "Kinetoscope Attachment for Stereoscopes," U.S. Patent No. 653,520 (July 10, 1900), 3.

39. C. Francis Jenkins, "Stereoscopic Mutoscope," U.S. Patent No. 671,111 (April 2, 1901), 2.

40. C. Francis Jenkins, "Device for Obtaining Stereoscopic Effects in Exhibiting Pictures," U.S. Patent No. 606,993 (July 5, 1898), 2.

41. C. Francis Jenkins, *Animated Pictures* (Washington, D.C.: By the Author, 1898), 94–100.

42. C. Francis Jenkins, "Stereoscopic Motion Pictures," *Transactions of the Society of Motion Picture Engineers* 9 (October 1920): 37–40.

43. C. Francis Jenkins, "History of the Motion Picture," *Transactions of the Society of Motion Picture Engineers* 9 (October 1920): 4–5.

44. G. W. Bitzer, *Billy Bitzer, His Story: The Autobiography of D. W. Griffith's Master Cameraman* (New York: Farrar, Straus Giroux, 1973), 10.

4. Stereoscopic Projection

1. Walter Stainton, "The Prophet Louis Ducos du Hauron and His Marvelous Moving Picture Machine," *Cinema Journal* 6 (1966–1967): 47.

2. H. Mark Gosser, *Selected Attempts at Stereoscopic Moving Pictures and Their Relationship to the Development of Motion Picture Technology, 1852–1903* (New York: Arno Press, 1977), 112.

3. Ibid., 113.

4. Louis Ducos du Hauron, Addition of December 3, 1864, to French patent no. 61,976 (March 1, 1864), 3.

5. Gosser, *Selected Attempts at Stereoscopic Moving Pictures,* 114.

6. Francois d'Aguilon, *Opticorum Libri Sex,* Book 6 (Antwerp: Plantin, 1613), 572.

7. Gosser, *Selected Attempts at Stereoscopic Moving Pictures,* 64.

8. Martin Quigley Jr., *Magic Shadows: The Story of the Origin of Motion Pictures* (New York: Quigley, 1960), 65.

9. Ibid.

10. E. J. Wall, "Stereoscopic Cinematography," *Transactions of the Society of Motion Picture Engineers* 10 (February 1927): 329.

11. William Van Doren Kelly, "Stereoscopic Pictures," *Transactions of the Society of Motion Picture Engineers* 17 (February 1924): 150.

12. Wilhelm Rollman, "Zwei neue stereoskopische Methoden," *Poggendorf's Annual* [*Annalen der Physik und Chemie*] 90 (August 1853): 186–87.

13. Joseph Charles d'Almeida, "Nouvel appareil stereoscopique," *Les Comptes Rendus de l'Academie des Sciences* 47 (July 12, 1858): 61–63, translated by Gosser, *Selected Attempts at Stereoscopic Moving Pictures*.

14. Ibid., 63, translated by Gosser, *Selected Attempts at Stereoscopic Moving Pictures*, 67.

15. G. Mareschal, "Projections Stereoscopiques," *La Nature* 917 (December 27, 1890): 49.

16. Letter from Louis Ducos du Hauron to M. Lelut of the Academie de Medecine et Sciences, 1862. This excerpt is from Louis Walton Sipley, *A Half Century of Color* (New York: Macmillan, 1951), 6–9.

17. Louis Ducos du Hauron, "Stereoscopic Print," U.S. Patent No. 544,666 (August 20, 1895), 2.

18. Ibid.

19. Arthur W. Judge, *Stereoscopic Photography*, 3rd ed., rev. (London: Chapman and Hall, 1950), 417.

20. John A. Norling, "The Stereoscopic Art: A Reprint," *Journal of the Society of Motion Picture and Television Engineers* 66 (March 1953): 297.

21. James L. Limbacher, *Four Aspects of the Film* (New York: Brussel and Brussel, 1969), 140.

22. Ray Allister, *Friese-Greene: Close-Up of an Inventor* (London: Marsland, 1948); this was a special reissue in 1951, including eight pages of photographs from the 1951 Festival of Britain film *The Magic Box,* which was based on the book, and a foreword by Robert Donat who portrayed Friese-Greene in the film.

23. Ibid., 184.

24. Ibid., xi.

25. Ibid., 45.

26. Brian Coe, "William Friese-Greene and the Origins of Cinematography III," *Screen: Journal of the Society for Education in Film and Television* 10 (July–October 1969): 132.

27. Allister, *Friese-Greene*, 46.

28. Gosser, *Selected Attempts at Stereoscopic Moving Pictures*, 183.

29. Coe, "William Friese-Greene," 138.

30. Correspondence, *Photographic Journal* 15 (November 21, 1890): 15.

31. Gosser, *Selected Attempts at Stereoscopic Moving Pictures*, 190.

32. Allister, *Friese-Greene*, 71.

33. F. Varley, "A Camera for Taking Consecutive Pictures of Objects in Motion," *British Journal of Photography* 37 (December 5, 1890): 777–80.

34. Allister, *Friese-Greene*, 130.

35. Coe, "William Friese-Greene," 140.

36. Ibid.

37. Ibid., 142.

38. This account is from Gosser (*Selected Attempts at Stereoscopic Moving Pic-

tures, 70), who cites the book Erasmus Bartholinus, *Experimenta crystalli Islandici disdiaclastici quibus mira et insolita refractio delegitur* (Hasniae: n.p., 1670).

39. John Anderton, "Method by Which Pictures Projected upon Screens by Magic Lanterns Are Seen in Relief," U.S. Patent No. 542,321 (July 9, 1895), 3.

40. Ibid., 3–4.

41. Gosser, *Selected Attempts at Stereoscopic Moving Pictures,* 247.

42. C. W. Ceram, *Archaeology of the Cinema* (New York: Harcourt, Brace and World, n.d.), 147.

43. Alexander Klein, "Skladanowsky," email to the author, June 29, 2005, citing Erich Skladanowsky, "60 Years Ago, A Newspaper Wrote . . .," *Image and Sound* [*Bild und Ton*] (November 1955).

44. Gosser, *Selected Attempts at Stereoscopic Moving Pictures,* 249.

45. Albert Narath, "Oskar Messter and His Work," *Journal of the Society of Motion Picture and Television Engineers* 69 (October 1960): 730.

46. This is *stereoskopischen Schnellseher* in German, which Alexander Klein translates as "fast stereoscopic vision device."

47. Narath, "Oskar Messter and His Work," 730.

48. Ibid., 731.

49. Mark Walker, *Ghostmasters: The Trials and Triumphs of America's Midnight Showmen* (n.p.: By the Author, 1991), 10.

50. Erik Barnouw, *The Magician and the Cinema* (New York: Oxford University Press, 1981), 5.

51. David Brewster, *Letters on Natural Magic: Addressed to Sir Walter Scott, Bart* (London: J. Murray, 1832; New York: Harper, 1842).

52. Henry Dircks, *The Ghost! As Produced in the Spectre Drama: Popularly Illustrating the Marvellous Optical Illusions Obtained by the Apparatus Called the Dircksian Phantasmagoria* (London: Spon, 1864).

53. The account of Theodore Brown and his work with the Kinoplastikon are from Stephen Herbert, *Theodore Brown's Magic Pictures: The Art and Inventions of a Multimedia Pioneer* (London: Projection Box, 1998), 80–87.

54. John Cher, "A Glimpse of Vienna and the Kinoplasticon," *Bioscope* (March 20, 1913), as cited in ibid., 82–83.

55. Theodore Brown, "Improvements in Cinematograph Apparatus for Producing Stereoscopic or Plastic Effects," British Patent No. 6,557 (1913).

56. *Times* (London), April 28, 1913, as cited in Herbert, *Theodore Brown's Magic Pictures,* 84.

57. Herbert, *Theodore Brown's Magic Pictures,* 86, 87.

58. Letter published in Malthete Lefebvre and Laurent Mannoni, *Lettres d'Etienne-Jules Marey a Georges Demeny, 1880–1894* (Paris: AFRHCBiBi, 2000), 495–96, translated by Laurent Mannoni, "The 'Feeling of Life': The Birth of Stereoscopic Film," in *Paris in 3D,* ed. Francoise Reynaud, Catherine Tambrun, and Kim Timby (Paris: Booth-Clibborn, 2000), 140. See also Mareschal, "Projections stereoscopiques."

59. Paul Mortier, "Appareil denomme Alethoscope, destine a enregistrer photographiquement les scenes animees et a les reproduire soit par projection, soit par vision directe avec ou sans L'illusion du relief," French Patent No. 254,090 (February 17, 1896).

60. Auguste Rateau, "Improvements in Chrono-Photographic Apparatus," British Patent No. 18,014 (July 13, 1897). The system was also discussed in *Le Moniteur de la photographie* 16 (August 15, 1898), as cited in Mannoni, "Feeling of Life," 142.

61. Claude Louis Grivolas, "Appareil pour projections animees en relief," French Patent No. 310,864 (May 20, 1901). These accounts of the work of Grivolas come from Gosser, *Selected Attempts at Stereoscopic Moving Pictures,* 270–74.

62. Gosser, *Selected Attempts at Stereoscopic Moving Pictures,* 250.

5. Cinema's Novelty Period

1. Charles Musser, *The Emergence of Cinema: The American Screen to 1907,* vol. 1 of *History of the American Cinema,* Charles Harpole, general editor (New York: Scribner's, 1990), 6.

2. Tom Gunning, "Now You See It, Now You Don't: The Temporality of the Cinema of Attractions," in *Silent Film,* ed. and with an introduction by Richard Abel (New Brunswick, N.J.: Rutgers University Press, 1996), 73. This influential essay first appeared in the magazine *From the Velvet Light Trap* 32 (1993), University of Texas Press.

3. C. Francis Jenkins, *Animated Pictures* (Washington, D.C.: H. L. McQueen, 1898), 5.

4. "New Magic Lantern Pictured," *Art-Journal* (London) (April 1851), 106.

5. Musser, *Emergence of Cinema,* 30.

6. Ibid., 31.

7. A. C. Wheeler, introduction to *A Guide to Cromwell's Stereopticon* (New York: n.p., ca. 1869), 8.

8. Musser, *Emergence of Cinema,* 38.

9. Gerry Turvey, "Panoramas, Parades and the Picturesque: The Aesthetics of British Actuality Films, 1895–1901," *Film History: An International Journal* 16, no. 1 (2004): 12.

10. Gunning, "Now You See It, Now You Don't," 75.

11. Maxim Gorky, "The Kingdom of Shadows," in *Authors on Film,* ed. Harry M. Geduld, trans. Leda Swan (Bloomington: Indiana University Press, 1972), 5. This was originally a review of the Lumiere program at the Nizhni-Novgorod Fair, as printed in the *Nizhegorodski listok* newspaper, July 4, 1896, and signed "I. M. Pacatus."

12. G. W. Bitzer, *Billy Bitzer, His Story: The Autobiography of D. W. Griffith's Master Cameraman* (New York: Farrar, Straus Giroux, 1973), 3–4.

13. Turvey, "Panoramas, Parades and the Picturesque," 9.

14. Ibid., 19.

15. Ibid., 20.

16. Cecil Hepworth, *Came the Dawn: Memories of a Film Pioneer* (London: Phoenix House, 1951), 44, 45.

17. Robert C. Allen, "Contra the Chaser Theory," in *Film before Griffith,* ed. John L. Fell (Berkeley: University of California Press, 1983), 109.

18. Quoted in ibid., 109, citing "Where the Past Speaks," *New York Mail and Express,* September 25, 1897.

19. Hepworth, *Came the Dawn,* 45.

20. *Edison Films,* Supplement no. 200, Edison Manufacturing Company, Orange, N.J., January Supplement, 1904, 5–7.

21. Tom Gunning, "An Unseen Energy Swallows Space: The Space in Early Film and Its Relation to American Avant-Garde Film," in *Film before Griffith,* ed. John L. Fell (Berkeley: University of California Press, 1983), 360.

22. Ibid., citing "The Spectator," *New York Dramatic Mirror* (April 19, 1910).

23. Noel Burch, *Life to Those Shadows,* trans. and ed. Ben Brewster (Berkeley: University of California Press, 1990), 6.

24. Ibid.

25. Ibid., 10.

26. Ibid., 162–63.

27. Ibid., 164.

28. Ibid., 173.

29. Giovanni Pastrone in an interview in Turin in 1949, cited in Georges Sadoul, *Histoire generale du cinema,* rev. ed. (Paris: Denoel, 1973), 215, translated in Burch, *Life to Those Shadows,* 181.

30. Originally published as Hugo Munsterberg, *The Photoplay: A Psychological Study* (New York: D. Appleton, 1916). These citations were taken from the reprint, retitled *The Film: A Psychological Study: The Silent Photoplay in 1916,* with new introduction by Richard Griffith (New York: Dover, 1970), 19.

31. Ibid., 20, 23, 30.

32. Willard Huntington Wright, "The Romance of the Third Dimension," *Photoplay Magazine* 20, no. 4 (September 1921): 41–42, quoted in George C. Pratt, ed., *Spellbound in Darkness: A History of the Silent Film,* 2nd ed. (New York: New York Graphic Society, 1973), 36.

33. Ibid.

34. Burch, *Life to Those Shadows,* 185 n.13, 184.

35. Munsterberg, *The Film,* 24.

36. Burch, *Life to Those Shadows,* 246.

37. Lynde Denig, "Stereoscopic Pictures Screened," *Moving Picture World* (June 26, 1915), 2072.

38. Lauren Kroiz, "The Reality of Vision with Depth: 3-D Cinematic Spectatorship," B.A. thesis, University of Chicago, 2002, 3. The word "scopophilia" is uncommon. It is derived from learned borrowings from the Greek, *skopion* "to

look" and *philia* (in pathological terminology), an abnormal liking for or tendency toward something. Also used with reference to voyeurism.

39. Jonathon Crary, *Techniques of the Observer: On Vision and Modernity in the Nineteenth Century* (Cambridge: MIT Press, 1990), 136.

40. Kroiz, "Reality of Vision with Depth," 30.

6. Public Exhibition of 3-D Films Begins

1. Lenny Lipton, *Foundations of the Stereoscopic Cinema: A Study in Depth* (New York: Van Nostrand Reinhold, 1982), 28.

2. H. Mark Gosser, *Selected Attempts at Stereoscopic Moving Pictures and Their Relationship to the Development of Motion Picture Technology, 1852–1903* (New York: Arno Press, 1977), 116.

3. Frederick Wenham, undated letter to the editor, *English Mechanic and the World of Science* 61 (June 7, 1895): 352.

4. Wilhelm Salow, "Stereoscopic Attachment for Photographic Cameras," U.S. Patent No. 840,378 (January 1, 1907), 2.

5. Aloys Wayditch, "Attachment for Taking Cameras for Making Stereoscopic Moving-Picture Films," U.S. Patent No. 1,071,837 (September 2, 1913), 3.

6. Aloys Wayditch, "Kinetoscope for Projecting Stereoscopic Moving Pictures," U.S. Patent No. 1,276,838 (August 27, 1918), 2.

7. Jules Richard and Louis Joseph Colardeau, "Stereoscopic Cinematograph," U.S. Patent No. 1,209,498 (December 19, 1916), 2.

8. A photo of the stereoscopic cinematograph is in Jacques Perin, "Stereoscopic Devices," in *Paris in 3D,* ed. Francoise Reynaud, Catherine Tambrun, and Kim Timby (Paris: Booth-Clibborn, 2000), 143.

9. Michael Sullivan, "Stereoscopic Projection Apparatus," U.S. Patent No. 1,189,308 (July 4, 1916), 2.

10. J. Henley, "Stereoscopic Projecting Machine," U.S. Patent No. 1,284,673 (November 12, 1918), 3.

11. E. I. Sponable, "Historical Development of Sound Films," *Society of Motion Picture Engineers Journal* 48 (April 1947): 280.

12. Frederick A. Talbot, *Moving Pictures: How They Are Made and Worked* (Philadelphia: Lippincott, 1912), 268.

13. Ibid.

14. Ibid., 272.

15. Pierre Mertz, "Historical Note: Lucien Georges Bull, 1876–1972," *Journal of the Society of Motion Picture and Television Engineers* 82 (August 1973): 683. Mertz also included a bibliography for additional research.

16. Ray Allister, *Friese-Greene: Close-Up of an Inventor* (London: Marsland, 1948), 67.

17. [By the English Correspondent,] "Stereoscopic Moving Pictures in Natural Colors," *Scientific American* (October 9, 1909), 256.

18. Ibid., 269.

19. Ibid.

20. William Friese-Greene, "Kinematographic Apparatus," U.S. Patent No. 937,367 (October 19, 1909), 2.

21. Bernard E. Jones, *How to Make and Operate Moving Pictures: A Complete Practical Guide to the Taking and Projecting of Cinematograph Pictures* (New York: Funk and Wagnalls, 1917), 185.

22. Allister, *Friese-Greene,* 147.

23. Terry Ramsaye, *A Million and One Nights: A History of the Motion Picture* (New York: Simon and Schuster, 1926), 571.

24. Ernest Betts, *The Film Business: A History of British Cinema, 1896–1972* (London: George Allen and Unwin, 1973), 30.

25. Roy Armes, *A Critical History of the British Cinema* (New York: Oxford University Press, 1978), 25.

26. Mathias J. Vinik, "Finder for Cameras," U.S. Patent No. 1,218,342 (March 6, 1917), 3.

27. North H. Losey, "Cinematography Apparatus," U.S. Patent No. 1,291,954 (January 21, 1919), 4.

28. Ibid.

29. Adolph Zukor with Dale Kramer, *The Public Is Never Wrong* (New York: Putnam's, 1953), 120.

30. Lynde Denig, "Stereoscopic Pictures Screened," *Moving Picture World* (June 26, 1915), 2072.

31. Ibid.

32. "Stereoscopic Films Shown," *New York Dramatic Mirror* (June 16, 1915).

33. Denig, "Stereoscopic Pictures Screened," 2072.

34. R. M. Hayes, *3-D Movies: A History and Filmography of Stereoscopic Cinema* (Jefferson, N.C.: McFarland, 1989), 3.

35. Zukor and Kramer, *The Public Is Never Wrong,* 120–21.

36. John B. Rathbun, *Motion Picture Making and Exhibiting* (Chicago: Charles C. Thompson, 1914), 230.

37. Frederic E. Ives, "Parallax Stereogram and Process of Making Same," U.S. Patent No. 725,567 (April 14, 1903), 2.

38. Herbert E. Ives, "Projection of Stereoscopic Pictures," U.S. Patent No. 1,883,290 (October 18, 1932).

7. A Wave of Stereo

1. Raymond Lecuyer (*Histoire de la photographie* [Paris: Baschet, 1945]) writes that printing inks up to the 1920s were not sufficiently precise and that available colors "were not strictly complementary and did not have the necessary vividness or transparency," as cited in Denis Pellerin, "The Anaglyph: A New Form of Stereoscopy," in *Paris in 3D,* ed. Francoise Reynaud, Catherine Tambrun, and Kim Timby (Paris: Booth-Clibborn Editions, 2000), 123.

2. Louis Ducos du Hauron, *L'Art des anaglyphes* (Algiers: Mustapha Printers, 1893), 4.

3. "The Explanation of Our Anaglyphs," *Illustrated London News* (March 8, 1921), 409.

4. Alfred J. Macy, "Stereoscopic Picture," U.S. Patent No. 1,386,720 (August 9, 1921), 2.

5. David Hutchison, "The Search for the Third Dimension," in *Fantastic 3-D: A Starlog Photo Guidebook*, ed. David Hutchison (New York: Starlog Press, 1982), 9.

6. Laurens Hammond, "Process of and Apparatus for Stereoscopic Shadow-graphs," U.S. Patent No. 1,481,006 (January 15, 1924), 2.

7. The Ohio Theater playbill with Follies-Scope glasses is in the author's collection.

8. "Screen: The Third Dimension," *New York Times* (October 22, 1922), sec. 8, p. 2.

9. "Three Dimension Motion Pictures Produced by Means of the Teleview," *Motion Picture News* (November 18, 1922), 2574.

10. Ibid.

11. Laurens Hammond, "Stereoscopic Picture Viewing Apparatus," U.S. Patent No. 1,658,439 (February 7, 1928), 3.

12. "The Screen: Vivid Pictures Startle," *New York Times* (December 28, 1922), 20.

13. Ibid.

14. Ibid.

15. "Stereoscopic Pictures Shown," *Film Daily* 21, no. 90 (September 30, 1922): 1.

16. "Movies Seen in Relief with Colored Glasses," *Popular Mechanics* 39 (April 1923): 513.

17. "Wearing Red and Green Spectacles to See Stereoscopic 'Movies,'" *Scientific American* 128 (February 1923): 105.

18. Harry K. Fairall, "Binocular Nonstop Motion Picture Camera," U.S. Patent No. 1,784,515 (December 9, 1930), 6.

19. "Wearing Red and Green Spectacles to See Stereoscopic 'Movies,'" 105; "Movies Seen in Relief with Colored Glasses," 513.

20. "Wearing Red and Green Spectacles to See Stereoscopic 'Movies.'"

21. "New Stereoscopic Camera for True Third Dimension," *New York Times* (February 20, 1927), 5.

22. William A. Crespinel, "Pioneer Days in Colour Motion Pictures with William T. Crespinel," *Film History: An International Journal* 12, no. 1 (2000): 67.

23. Ibid.

24. Ibid., 68.

25. Ibid.

26. Charles Raleigh and William V. D. Kelley, "Producing Colored Photographic Pictures," U.S. Patent No. 1,217,425 (February 27, 1917).

27. W. T. Crespinel, "Color Photography: Yesterday, Today and Tomorrow," *American Cinematographer* (March 1929), 4.

28. William Van Doren Kelley, "Natural Color Cinematography," *Society of Motion Picture Engineers Journal* (November 1918): 38.

29. Leon F. Douglass, "Cinematography," U.S. Patent No. 1,313,587 (August 19, 1919), 3; Leon F. Douglass, "Method of an Apparatus for Producing Stereoscopic Photographic Films," U.S. Patent No. 1,429,495 (September 19, 1922).

30. William Van Doren Kelley, "Stereoscopic Pictures," *Society of Motion Picture Engineers Journal* 17 (October 1923): 151.

31. Ibid.

32. William Van Doren Kelley and Dominick Tronolone, "Stereoscopic Picture," U.S. Patent No. 1,729,617 (October 1, 1929), 3, 6.

33. Titled *Thru the Trees* and crediting photography to William Crespinel, this short anaglyph film was screened, as twin-strip black and white with dual polarizing projection, September 17, 2006, at the World 3D Film Expo II at the Egyptian Theater in Hollywood. It was found in early 2006 by film historian Jack Theakston at the Library of Congress and, under the auspices of Jeff Joseph and Sabucat Productions, 3-D filmmaker and historian Dan Symmes restored the film, extracting the left-right pairs as black and white from the anaglyph print. On his website, Symmes tells the full story of the restoration of this important stereoscopic motion picture and provides a highly detailed history behind it. The website is http://www.3dmovingpictures.com/.

34. Kelley, "Stereoscopic Pictures," 152.

35. J. A. Norling and J. F. Leventhal, "Some Developments in the Production of Animated Drawings," *Society of Motion Picture Engineers Journal* (September 1926): 62–63.

36. The Paramount Logo Card at the opening of a 1930s Fleischer Brothers Superman cartoon bore the legend "Stereoptical Process and Apparatus Patented, Patent Number 2,054,414." The U.S. patent in question, issued to Max Fleischer, was dated September 15, 1936, and titled "Art of Making Motion Picture Cartoons." The patent drawing depicted a multiplane filming stage with cells and a physical background, which would move horizontally so that "the objects of the background will appear to stand out in bas-relief" (3).

37. Crespinel, "Pioneer Days in Colour Motion Pictures," 65.

38. Ibid.

39. Ibid.

40. Ibid., 66.

41. "'Third Dimension' Effect Realized," *Motion Picture News* (January 23, 1924), 875.

42. "Educational Augments March List," *Motion Picture News* (March 8, 1924), 1095. The diversity of the output from Educational Films Corp. is evident in the following listing of releases for March 1924 (in addition to *Plastigrams*): "The two-reel subjects include two Mermaid Comedies, two Christie Comedies, and one

each of the Tuxedo, Clyde Cook, and Juvenile Comedies, while the seven single reel subjects include three Cameo Comedies, and one subject each of the Bruce Wilderness Tales, Secrets of Life series, Lyman H. Howe's Hodg-Podge and the Sing Them Again series. Ten issues of Kinograms, issued twice weekly, complete the month's program."

43. Earl W. Hammons, "Why Not a Short Subject Theatre?" *Motion Picture News* (March 15, 1924), 1177.

44. Quoted in "'Third Dimension' Effect Realized," 875.

45. Ibid.

46. *Plastigrams* advertisement in *Motion Picture News* (March 15, 1924), 1157.

47. "New 'Plastigrams' Are Widely Booked," *Motion Picture News* (March 22, 1924), 1333.

48. "New 'Third Dimension' Series for Pathe," *Motion Picture News* (February 4, 1925), 701.

49. Crespinel, "Pioneer Days in Colour Motion Pictures," 66.

50. R. M. Hayes, *3-D Movies: A History and Filmography of Stereoscopic Cinema* (Jefferson, N.C.: McFarland, 1989), 9.

51. Maurice H. Zouary, *DeForest: Father of the Electronic Revolution* (n.p.: By the Author, 2000), 126.

52. Ibid., 128.

53. Advertisements, *Motion Picture News* (March 15, 1925) and (March 22, 1925).

54. All of these anaglyphic vignettes were shown with a 35-mm print provided by Eastman House at the first World 3D Film Expo on September 20, 2003, and again at World 3D Film Expo II on September 17, 2006.

55. J. F. Leventhal, "The First Use of Stereoscopic Pictures in Motion Picture Theatres," *Society of Motion Picture Engineers Journal* (November 1926): 34.

56. Ibid., 35, 36.

57. Ibid., 36.

58. Crespinel, "Pioneer Days in Colour Motion Pictures," 66.

8. Essaying Utopia

1. Fred N. Hallett, "Moving-Picture Camera," U.S. Patent No. 1,363,249 (December 28, 1920), 2.

2. Clement A. Clement and Axel Bors-Koefoed, of Houston, Texas, said Bors-Koefoed Assignor to said Clement, Trustee, "Motion Picture Machine," U.S. Patent No. 1,477,541 (December 18, 1923), 3.

3. Edgar Blackburn Moore, "Stereoscopic-Motion-Picture Mechanism," U.S. Patent No. 1,396,651 (November 8, 1921), 2.

4. Raymond A. Duhem and John D. Grant, "Method of Producing Stereoscopic Effects in Motion Pictures," U.S. Patent No. 1,351,508 (August 31, 1920), 4.

5. Waldon S. Ball, Assignor to Stereoscopic Motion Picture Company, "Method of and Apparatus for Taking Moving Pictures," U.S. Patent No. 1,351,502 (August 31, 1920), 3.

6. Stephen Herbert, *Theodore Brown's Magic Pictures: The Art and Inventions of a Multimedia Pioneer* (London: Projection Box, 1998), 50.

7. Theodore Brown, writing in *Optician* (October 28, 1904), as quoted in Herbert, *Theodore Brown's Magic Pictures,* 55.

8. "Three Dimension Photography Is Perfected: New Camera Will Make It Possible to Project Dimension of Depth on Screen, Is Claim," *Motion Picture News* (January 22, 1921), 828.

9. Edward Connor, "3-D on the Screen: Its Potentialities Could Save the Lives of Many Movie Theatres," *Films in Review* (March 1966), 161.

10. "Spoor Starts Production of 'Natural Vision' Film," *Motion Picture News* (April 18, 1925), 725.

11. As quoted in Connor, "3-D on the Screen," 162.

12. As quoted in Kemp Niver and Arthur Miller, audiocassette found at the American Society of Cinematographer's Clubhouse, dated September 1963.

13. As quoted in "Screen," *New York Times* (January 23, 1921), sec. 6, p. 2.

14. Ibid.

15. Connor, "3-D on the Screen," 161.

16. "Screen," *New York Times* (May 15, 1923), sec. 2, p. 22.

17. As quoted in Connor, "3-D on the Screen," 163.

18. Ibid.

19. Lynde Denig, "Stereoscopic Pictures Screened," *Moving Picture World* (June 26, 1915), 2072.

20. Maurice H. Zouary, *DeForest: Father of the Electronic Revolution* (n.p.: By the Author, 2000), 126–27.

21. Ibid., 127.

22. "Size of Screen," *New York Times* (June 1, 1930), sec. 9, p. 6.

23. Ibid.

24. *New York Times* (May 25, 1929), sec. 8, p. 16.

25. Abel Gance, *Napoleon,* trans. Moya Hassan, ed. Bambi Ballard (London: Faber and Faber, 1990), xxi, xxii.

26. Quoted in Kevin Brownlow, *The Parade's Gone By* (Berkeley: University of California Press, 1968), 559.

27. Ibid.

9. Stereoscopic Cinema Proves Itself

1. R. M. Hayes, *3-D Movies: A History and Filmography of Stereoscopic Cinema* (Jefferson, N.C.: McFarland, 1989), 132.

2. James L. Limbacher, *Four Aspects of the Film* (New York: Brussel and Brussel, 1969), 140.

3. Auguste Lumiere and Louis Lumiere, *Letters: Inventing the Cinema* (London: Faber and Faber, 1994), 142, 314.

4. "Disc Stereo Device," French Patent No. 305,092 (November 3, 1900), 2, translated by H. Mark Gosser, *Selected Attempts at Stereoscopic Moving Pictures and Their Relationship to the Development of Motion Picture Technology, 1852–1903* (New York: Arno Press, 1977), 284.

5. Gosser, *Selected Attempts at Stereoscopic Moving Pictures*, 285.

6. Louis Lumiere, "Stereoscopy on the Screen," *Society of Motion Picture Engineers Journal* 27 (September 1936): 315, 318.

7. Ibid., 317.

8. Francoise Reynaud, Catherine Tambrun, and Kim Timby, eds., *Paris in 3D* (Paris: Booth-Clibborn, 2000), 123.

9. Georges Sadoul, "Lumiere: The Last Interview," *Sight and Sound* 17 (January 1948): 70.

10. Frank Nugent, "Meet the Audioscopiks: Metro's New Three-Dimensional Film Is the Next Novelty on the Schedule," *New York Times* (December 8, 1935), sec. 5, p. 6.

11. Scott Eyman, *The Speed of Sound: Hollywood and the Talkie Revolution, 1926–1930* (Baltimore: Johns Hopkins University Press, 1997), 243.

12. During his first year as head of publicity at MGM, Pete Smith had an agonizing experience with sound broadcast to theaters via radio to run with a trailer for the feature film *Slave of Fashion*. Produced at the behest of Douglas Shearer, the sound was horribly out of sync with the trailer as it ran in the theaters (related in Bosley Crowther, *The Lion's Share: The Story of an Entertainment Empire* [New York: E. P. Dutton, 1957], 138–39).

13. J. A. Norling, "Three-Dimensional Motion Pictures," *Society of Motion Picture Engineers Journal* (December 1939): 612.

14. Nugent, "Meet the Audioscopiks."

15. H. T. Kalmus, "Technicolor Adventures in Cinemaland," *Society of Motion Picture Engineers Journal* 31 (December 1938), as quoted in Raymond Fielding, ed., *A Technological History of Motion Pictures and Television* (Berkeley: University of California Press, 1967), 53.

16. Ibid., 54, 55.

17. Pete Smith, "Three Dimensionally Speaking," in *New Screen Techniques,* ed. Martin Quigley (New York: Quigley, 1953), 17.

18. Ibid., 18.

19. Ibid., 19.

20. Norling, "Three-Dimensional Motion Pictures," 617.

21. Ibid.

22. Victor K. McElheny, *Insisting on the Impossible: The Life of Edwin Land, Inventor of Instant Photography* (Reading, Mass.: Perseus, 1998), 33.

23. Edwin H. Land, "Polarizing Refracting Bodies," U.S. Patent No. 1,918,848 (July 18, 1933), 2.

24. Quoted in McElheny, *Insisting on the Impossible,* 111.

25. Clarence Kennedy, "The Development and Use of Stereo Photography for Educational Purposes," *Society of Motion Picture Engineers Journal* 26 (January 1936): 3.

26. William H. Ryan, "Polaroid and 3-D Films," in *New Screen Techniques,* ed. Martin Quigley (New York: Quigley, 1953), 28.

27. Edwin H. Land, "Polarizing Optical System," U.S. Patent No. 2,099,694 (November 23, 1937), 1.

28. Waldemar Kaempffert, "Three-Dimension Movies," *New York Times* (February 16, 1936), 6.

29. *New York Herald Tribune* (May 14, 1936), quoted in McElheny, *Insisting on the Impossible,* 114.

30. "Death after Dark: Device to Cut Toll, Headlight Glare Dimmed; Three-Dimension Movies Arrive," *Literary Digest* (December 12, 1936), 30.

31. Frank A. Weber, "3-D in Europe," in *New Screen Techniques,* ed. Martin Quigley (New York: Quigley, 1953), 73, 71.

32. Eddie Sammons, *The World of 3-D Movies* (n.p.: A Delphi Publication, 1992), 130.

33. *Cinema* (December 10, 1936), as quoted in ibid.

34. Dieter Lorenz, "Zeiss Ikon and Stereo Cinematography," *Stereo World* 31, no. 1 (2004–2005): 7. Available from the National Stereoscopic Association, Portland.

35. Norling, "Three Dimensional Motion Pictures," 628. The Technicolor version of this film, retitled *Motor Rhythm* by RKO for 3-D release in 1953, was screened on September 12, 2003, at the Egyptian Theater in Hollywood for the World 3D Film Expo, and again on September 17, 2006, for World 3D Film Expo II.

36. Richard Griffith, "Films at the Fair," *Films* 1, no. 1 (November 1939): 72.

37. Norling, "Three Dimensional Motion Pictures," 628.

38. John A. Norling, "The Stereoscopic Art: A Reprint," *Journal of the Society of Motion Picture and Television Engineers* 60 (March 1953): 289.

39. A newly found print of *Thrills for You* was restored by Jeff Joseph and Daniel Symmes and screened at the World 3D Film Expo II at the Egyptian Theater in Hollywood on September 17, 2006.

40. J. A. Norling, "Progress in Three-Dimensional Pictures," *Society of Motion Picture Engineers Journal* 37 (November 1941): 516.

41. Richard T. Kriebel, "Stereoscopic Photography," *Complete Photographer* 53 (1943): 3463, as quoted in McElheny, *Insisting on the Impossible*, 116.

42. J. A. Norling, "Light Control by Polarization and the Application of Polarizers to the Stereoscopic Process," *Journal of the Society of Motion Picture and Television Engineers* 48, no. 2 (February 1947), 142.

43. Edwin H. Land and Joseph Mahler, "Apparatus Employing Polarized Light for the Production of Stereoscopic Images," U.S. Patent No. 2,203,687 (June 11, 1940), 3.

44. Quoted in McElheny, *Insisting on the Impossible*, 116.

45. Edwin H. Land, "Vectographs: Images in Terms of Vectorial Inequality and Their Applications in Three-Dimensional Representation," *Journal of the Optical Society of America* 30, no. 6 (June 1940): 230.

46. Josef Mahler, "Stereo Apparatus," U.S. Patent No. 1,992,872 (February 26, 1935), 3.

47. Joseph Mahler, "Apparatus for Viewing Stereoscopic Pictures," U.S. Patent No. 2,574,186 (November 6, 1951), 2.

48. Vivian Walworth, interview with the author, San Jose, California, January 17, 2006. Walworth worked with Edwin Land at Polaroid in the postwar years developing Vectograph motion pictures.

49. This circular is in the possession of the author.

10. The Stereoscopic Overture Finishes

1. "New Screen Gives Depth to Movies," *New York Times* (November 24, 1931), sec. 1, p. 27.

2. Ibid.

3. Frederic E. Ives, "Parallax Stereogram and Process of Making Same," U.S. Patent No. 725,567 (April 14, 1903).

4. John Jacobson, "Pictorial Reproduction," U.S. Patent No. 624,042 (May 2, 1899).

5. "Movies in Relief Forecast by Test," *New York Times* (October 31, 1930), sec. 1, p. 19.

6. Herbert E. Ives, "The Problem of Projecting Motion Pictures in Relief," *Society of Motion Picture Engineers Journal* (April 1932): 437.

7. "Movies in 'Depth' Achieved at Last," *New York Times* (April 28, 1933), sec. 2, p. 25.

8. Gregg Toland, "Camera in Revolt: Three-Dimensional Photography Heralds the Dawn of a New Era in Pictures," *New York Times* (January 20, 1935), sec. 10, p. 5.

9. Ibid.

10. Ibid.

11. Patrick L. Ogle, "Technological and Aesthetic Influences upon the Development of Deep Focus Cinematography in the United States," *Screen: Journal of the Society for Education in Film and Television* 13 (Spring 1972): 45.

12. H. Mark Gosser, *Selected Attempts at Stereoscopic Moving Pictures and Their Relationship to the Development of Motion Picture Technology, 1852–1903* (New York: Arno Press, 1977), 302.

13. Adrian Cornwell-Clyne, *3-D Kinematography and New Screen Techniques* (London: Hutchisons, 1954), 72.

14. S. Ivanov, "Russia's Third Dimensional Movies," *American Cinematographer* (May 1941), 212.

15. Ibid., 213.

16. Ivor Montagu, *Film World: A Guide to Cinema* (Middlesex: Penguin, 1964), 91.

17. Joseph MacLeod, "Stereoscopic Film: An Eye Witness Account," *Sight and Sound* 16 (Autumn 1947): 118.

18. S. M. Eisenstein, "About Stereoscopic Cinema," trans. Catherine de la Roche, *Penguin Film Review* 8 (January 1949): 35, as quoted in *Film Makers on Film Making,* ed. Harry Geduld (Bloomington: Indiana University Press, 1967), 106.

19. Frank A. Weber, "Stereofilm Making with the VeriVision Camera," *American Cinematographer* (May 1952), 204.

20. Frank A. Weber, "3-D in Europe," in *New Screen Techniques,* ed. Martin Quigley (New York: Quigley, 1953), 70.

21. Weber, "Stereofilm Making with the VeriVision Camera," 220.

22. Ibid.

23. Earle F. Watts and John T. Rule, *Descriptive Geometry* (New York: Prentice Hall, 1946); John T. Rule and Earle F. Watts, *Engineering Graphics* (New York: McGraw-Hill, 1951).

24. John T. Rule, "Stereoscopic Drawings," *Journal of the Optical Society of America* 28 (August 1938): 313.

25. John T. Rule, "The Shape of Stereoscopic Images," *Journal of the Optical Society of America* 31 (February 1941): 124.

26. John T. Rule, "The Geometry of Stereoscopic Projection," *Journal of the Optical Society of America* 31 (April 1941): 325.

27. Ibid., 334.

28. H. Vuibert, *Les Anaglyphes geometriques* (Paris: Librarie Vuibert, 1912).

29. Man Ray, *Self-Portrait* (New York: Little, Brown, 1979), 101–2.

30. Ibid., 275.

31. William Moritz, "The Films of Oskar Fischinger," *Film Culture* 58–60 (1974), 72.

32. Ibid.

33. William Moritz, *Optical Poetry: The Life and Work of Oskar Fischinger* (Eastleigh, Eng.: John Libby, 2004), 166, 167.

34. Raymond Spottiswoode, "Progress in Three-Dimensional Films at the Festival of Britain," *Journal of the Society of Motion Picture and Television Engineers* 58 (April 1952): 298.

35. Raymond Spottiswoode and Nigel Spottiswoode, *The Theory of Stereoscopic Transmission and Its Application to the Motion Picture* (Berkeley: University of California Press, 1953).

36. Spottiswood, "Progress in Three-Dimensional Films," 297.

37. Ibid.

38. Norman Jenkins, "The Cash Customers at the Festival of Britain's Telecinema," *Society of Motion Picture Engineers Journal* 58 (April 1952): 307.

39. Norman McLaren, "Stereographic Animation: The Synthesis of Stereo-

scopic Depth from Flat Drawings and Art Work," *Society of Motion Picture Engineers Journal* 57 (December 1951): 317, 313.

40. Gerald Pratley, "The Latest 3-Dimensional Films Prove That the Movies Still Have an Ace up Their Sleeve," *Films in Review* (May 1952), 171.

Epilogue

1. R. M. Hayes, *3-D Movies: A History and Filmography of Stereoscopic Cinema* (Jefferson, N.C.: McFarland, 1989), 15.

2. Raymond Spottiswoode, "Progress in Three-Dimensional Films at the Festival of Britain," *Journal of the Society of Motion Picture and Television Engineers* 58 (April 1952): 300.

3. *The Diamond Wizard*, released July 16, 1954, was initially distributed in "flat" 2-D versions only. The stereoscopic version was restored by Jeff Joseph, Daniel Symmes, Bob Furmanek, and Sabucat Productions and given its world premiere in 3-D on September 13, 2006, at the World 3D Film Expo II in Hollywood.

4. Hayes, *3-D Movies,* 19.

5. Raymond Spottiswoode, N. L. Spottiswoode, and Charles Smith, "Basic Principles of the Three-Dimensional Film," *Journal of the Society of Motion Picture and Television Engineers* 59 (October 1952): 249. The parenthetical usage of the term "3-D" occurs in the first sentence of the paper when the authors write, "Up to now the production of three-dimensional (3-D) films has been sporadic—scattered all over the world and separated by long intervals of time."

6. Raymond Spottiswoode and Nigel Spottiswoode, *The Theory of Stereoscopic Transmission and Its Application to the Motion Picture* (Berkeley: University of California Press, 1953), vii.

7. Ibid., 156.

8. Ray Zone, "Behind the Scenes of *Bwana Devil*," *Stereo World* 29, no. 2 (2002): 11. Portions of the Lothrop Worth interview with Mike Hyatt are included in this article.

9. Andrew Dowdy, *The Films of the Fifties: The American State of Mind* (New York: William Morrow, 1973), 47.

10. Zone, "Behind the Scenes," 10.

11. Victor K. McElheny, *Insisting on the Impossible: The Life of Edwin Land, Inventor of Instant Photography* (Reading, Mass.: Perseus, 1998), 128.

12. Letter from John T. Rule to John Norling, dated March 25, 1952, typed on MIT letterhead, in possession of the author.

13. Hectograph of the Society of Motion Picture and Television Engineers, memo 591, dated March 19, 1953, in possession of the author.

14. Azograph drawing in possession of the author.

15. Hectograph of the Society of Motion Picture and Television Engineers, memo 591, 8.

16. William Paul, "Breaking the Fourth Wall: 'Belascoism,' Modernism, and a 3-D *Kiss Me Kate*," *Film History: An International Journal* 16, no. 3 (2004): 229.

17. Spottiswoode and Spottiswoode, *Theory of Stereoscopic Transmission*, 156.

18. S. M. Eisenstein, "About Stereoscopic Cinema," trans. Catherine de la Roche, *Penguin Film Review* 8 (January 1949), as quoted in *Film Makers on Film Making*, ed. Harry Geduld (Bloomington: Indiana University Press, 1967), 114.

Index